NETWORK INTELLIGENCE

BT Telecommunications Series

The BT Telecommunications Series covers the broad spectrum of telecommunications technology. Volumes are the result of research and development carried out, or funded by, BT, and represent the latest advances in the field.

The series includes volumes on underlying technologies as well as telecommunications. These books will be essential reading for those in research and development in telecommunications, in electronics and in computer science.

NETWORK INTELLIGENCE

Edited by

I.G. Dufour
BT Laboratories
Martlesham Heath
UK

CHAPMAN & HALL

London · Weinheim · New York · Tokyo · Melbourne · Madras

Published by Chapman & Hall, 2–6 Boundary Row, London SE1 8HN, UK

Chapman & Hall, 2–6 Boundary Row, London SE1 8HN, UK

Chapman & Hall GmbH, Pappelallee 3, 69469 Weinheim, Germany

Chapman & Hall USA, 115 Fifth Avenue, New York, NY 10003, USA

Chapman & Hall Japan, ITP-Japan, Kyowa Building, 3F, 2-2-1 Hirakawacho, Chiyoda-ku, Tokyo 102, Japan

Chapman & Hall Australia, 102 Dodds Street, South Melbourne, Victoria 3205, Australia

Chapman & Hall India, R. Seshadri, 32 Second Main Road, CIT East, Madras 600 035, India

First edition 1997

© 1997 British Telecommunications plc

Printed in Great Britain by T.J. Press Ltd, Padstow, Cornwall

ISBN 0 412 78920 5

A catalogue record for this book is available from the British Library

∞ Printed on permanent acid-free text paper, manufactured in accordance with ANSI/NISO Z39.48-1992 and ANSI/NISO Z39.48-1984 (Permanence of Paper).

Contents

Contributors

T W Abernethy	Intelligent Networks, BT Laboratories,
M Bagley	Applications Research, BT Laboratories,
M C Bale	Signalling and Protocols, BT Laboratories,
R G Buck	Integrated Services, BT Laboratories,
D R Chesterman	Switching Systems, BT Laboratories,
I G Dufour	Network Intelligence, BT Laboratories,
P E Holmes	Service Management, BT Laboratories,
P M Hughes	Speech Systems, BT Laboratories,
S Kabay	Formerly Service Control Development, BT Laboratories,
N C Lobley	Cellular Systems, BT Laboratories,
I W Marshall	Applications Research, BT Laboratories,
P A Martin	Network Intelligence, BT Laboratories,
A C Munday	Intelligent Networks, BT Laboratories,
K R Rose	Speech Systems, BT Laboratories,
C J Sage	Formerly Service Control Development, BT Laboratories,
D G Smith	Cellular Systems, BT Laboratories,
N J Street	Service Management, BT Laboratories,
R P Swale	Core Platforms, BT Laboratories,
G D Turner	Service Development, BT Laboratories,
P Willis	Formerly Internet Protocol Engineering, BT Laboratories,
K C Woollard	Signalling and Protocols, BT Laboratories,

Preface

An intelligent network means different things to different people. Some interpret it widely as a network that finds you, talks to you and knows what you want. Some see it in terms of the separation of service control logic from core-network routeing logic in the manner encapsulated by the various standards bodies — perhaps the most conventional interpretation of the intelligent network (IN). Others see it as the convergence of computing and telecommunications, sometimes described as computer-driven telecommunications.

Little is certain about an IN — it is an open-ended concept and is, in practice, little more than a design philosophy for a flexible network moving into the information age. The attraction of the term network intelligence, which is the theme of this book, is that it embraces all these views and more, including the movement away from fixed-speech networks to those based on datacommunications principles, the trend to greater mobility and the synergies and trade-offs between core-network intelligence and intelligent appliances at the periphery leading to the application of distributed processing principles.

This book has two strands. The first eight chapters cover the more conventional IN aspects associated with speech-based core networks. These chapters bring together many of the issues being addessed now by operators throughout the world, as well as BT, in implementing an advanced IN. The second strand, comprising a further six chapters, explores the boundaries of network intelligence beyond IN.

The book commences with an overview of intelligent networks, standards and services. The origins of IN are covered, together with a comprehensive outline of the design principles lying behind the standards which have been put in place by the international standards bodies. Some typical realizations are also outlined. Chapter 2 describes cellular radio networks which are growing in size and importance. Cellular networks have been leaders in the application of intelligence — it is mainly used at present within these networks to track the location of the user. However, design principles are converging with those of the fixed IN, such that the third-generation cellular networks described will embrace IN principles. The stage will soon be reached where overall network design allows for total mobility and, if there is any subsequent distinction between fixed and mobile networks other than in using radio for access, it will mainly be for commercial or regulatory reasons.

Chapter 3 deals with the important issue of signalling systems. INs require huge amounts of signalling information to be passed between network elements. While the structure of Signalling System No 7 is already well documented, little has hitherto been published about the IN application prototcol described here. The following chapter demonstrates that the networks and complex signalling systems described in the first three chapters raise many issues of service definition and interworking and goes on to outline the use of modelling and formal methods for the design and specification of IN-based services.

Many of the advanced services now being delivered from the IN provide challenges at the network-to-customer interface. Speech systems are fundamental to the operation of many of these and future services, and are described in Chapter 5 which provides an overview of the subject before looking at some IN-specific issues. Chapter 6 addresses the architectural and technical issues identified during the development of a service node where many of the key aspects of network control, signalling systems, computer processing, speech systems, service interworking and systems integration come together.

To provide a usable service to customers the network systems described so far have to be managed. This is the role of operational support systems which are described in Chapter 7 which puts them in the context of an IN, showing that a flexible network calls for flexible management systems which pose special challenges for their designers. Chapter 8 concludes the main-stream IN activities with a review of service creation. If an understanding that IN is just a synonym for a flexible network has taken some years to be formed, a full understanding of service creation has some years yet to mature. Service creation addresses business processes which are at the heart of a telco operation and where the provision of a new service requires the alignment not just of network systems but of the operational support systems and the delivery chain to the customer as well.

The second strand of this book looks at a wider definition of intelligence and at how the underlying telecommunications and computing technologies are influencing developments more broadly. It starts with the view that the intelligent network, using the widest definition, will possess knowledge of the user and may one day more properly be called a knowledge-based network. Chapter 10 develops part of this theme by describing intelligence in terms of its relationship with the applications embodied within both public and private networks and in particular shows the relationships between IN and computer/telephony integration (CTI). Chapters 11 and 12 look at how a new generation of terminals will put users in control of a wide variety of new services, spanning fixed and mobile networks and speech and non-speech services, and at the opportunities to use ISDN signalling in conjunction with screen phones to enhance the usability of the IN.

Without exception the new era of telecommunications outlined here will be dependent on software, and the penultimate chapter notes that some key developments in distribution and object-oriented techniques will be needed to support the high-bandwidth, low-delay information flows of future networks. The book concludes with a case study of a prototype which draws together many of the trends outlined in the preceding chapters.

BT is at the forefront of major developments in the dynamic and global telecommunications industry. Much of this stems from the work carried out at BT Laboratories which ranges from long-term research to the detailed design and implementation of operational systems. In putting this book together I have called for contributions from numerous leaders in their field from the Laboratories. I am very grateful for the work they have put in and the time they have given to the task. I am pleased with the result — I hope you are too. I and all the contributors also recognise the tremendous support we get from our numerous colleagues both at BT Laboratories and throughout BT as a whole, many of whom have contributed to the work described here. I at least, but possibly not the authors, also hope that this will lead to a further book expanding on some of the issues raised in the later chapters.

I would like to end this preface on an historical note. All the papers in this edition are concerned with various aspects of the convergence of telecommunications and computing and it is worth remembering that the world's first programmable electronic digital computer was built in a telecommunications laboratory. The authors of these papers are building on the foundations laid by their predecessors, then based at the General Post Office Research Station, Dollis Hill, London. The computer they built, known as COLOSSUS, was designed to decypher German codes — a completely different variety of intelligence — during the Second World War and first came into use at Bletchley Park (where a replica can be found in the museum there) in February 1944. The team was led by Dr T H Flowers who is a leading figure in the history of computing and who therefore should also be recognised as contributing to the origins of network intelligence.

Ian Dufour
Network Intelligence Engineering Centre Manager
BT Laboratories

e-mail:dufouri@boat.bt.com
BT Laboratories website: http://www.labs.bt.com

1

INTELLIGENT NETWORKS, STANDARDS AND SERVICES

T W Abernethy and A C Munday

1.1 INTRODUCTION

Over the last three decades, there have been a number of significant technological developments in public switched networks (PSTN), namely stored programme control switches, digital switching, ISDN and recently the introduction of intelligent networks (INs).

It takes several years to specify, develop, test and deploy new switch-based services, with additional features waiting for a time-slot in the switch development cycle. Also, the time taken to upgrade a large network with new software, and perhaps hardware, can exceed a year, constituting a considerable time lag from conception of a new service to the first customer using the service.

Hence, the major drivers for the addition of intelligence to the network are the:

- requirement for the rapid introduction of new services and features;

- reduction of the reliance on switch manufacturers for the provision of new services;

- ability to offer integrated service packages.

Through the deployment of intelligent networks telecommunications operators are endeavouring to meet the demands of these drivers.

The intelligent network development is essentially humanizing the network — adding to the network the intimate knowledge of the user and those connected to it — taking the network 'back' to the days when Strowger thought the network too 'intelligent' and developed the automatic switch.

The following sections describe the current standardization activities, the IN conceptual model and the IN components found in the network. Also included is an overview of some of the services implemented using an intelligent network. Some of the developments occurring in the field of intelligent networks are also outlined.

1.2   INTELLIGENT NETWORK STANDARDS

A major step forward in intelligent networks was made when the Regional Bell Operating Companies (RBOCs) asked Bellcore to produce the advanced intelligent network (AIN) series of technical requirements. The development of AIN followed a series of multi-vendor discussions which took place during the late 1980s and which resulted in the production of AIN Release 0.0, leading on to the development of AIN Release 0.1 — with AIN Release 1.0 being the target AIN architecture for the next decade. Other standards bodies, for example, the International Telecommunications Union — Telecommunications Standardization Sector (ITU-T) and European Telecommunications Standards Institute (ETSI), have now also produced definitions for the intelligent network architecture using a range of conceptual and functional models (see section 1.2.1). Figure 1.1 shows the relationships between the USA, European and international standards development process.

Figure 1.2 illustrates the ITU-T processes for the production of IN capability set 1 (CS-1), which forms part of the input to the definition of IN CS-2 and 3 along with contributions from ETSI in the form of 'Core-INAP' (intelligent network application protocol).

The IN CS-1 recommendations are in the Q.121x series of documents, with IN CS-2 contained in Q.122x. These documents are developments from the base Q.1200 series of recommendations.

The Q.12xx recommendations provide the definition of the IN architecture through the use of a conceptual model.

1.2.1   The IN conceptual model

The IN concepts can be explained in terms of a model which provides a framework for the design and description of the IN architecture. This IN conceptual model (INCM) has been used as a tool during the development of the current IN capability sets. By explaining this model the IN concepts can be understood.

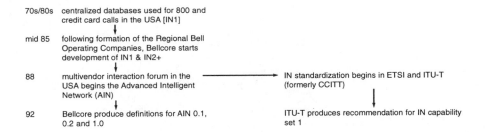

Fig 1.1   History of the development of AIN and IN capability set standards.

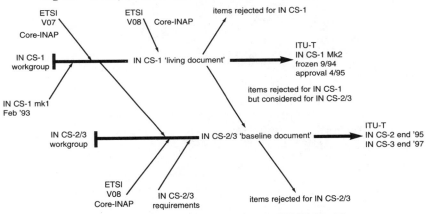

Fig 1.2   ITU-T process for the production of IN CS-2 and 3.

/The IN conceptual model consists of four planes each of which provides a different abstract view of the capabilities of an IN structured network. This ranges from how services are defined, at the top layer, to how they are mapped to a physical network, the bottom layer (see Fig. 1.3).

The INCM can be used to provide both a top-down view with respect to services and a bottom-up view with respect to networks.

1.2.1.1   The service plane

The service plane provides a service-oriented view and describes services in terms of service features. The service features are themselves described in terms of global service logic. Each service feature is defined by a single set of global service logic (GSL). At this high level there is no view of the relationship of the service or service feature with the physical network on which it is deployed.

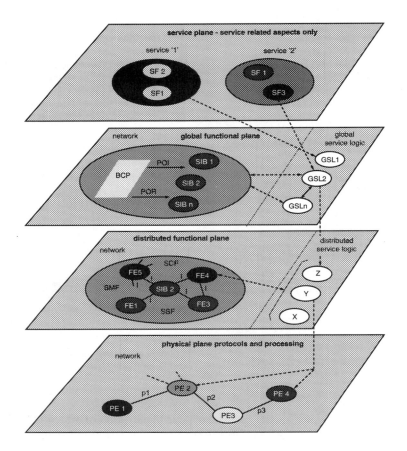

Fig. 1.3   The IN conceptual model (based on Q.1201 [1]).

In a high-level service-creation environment something representing the service feature level of granularity for creating new services would be expected. How fine or course this granularity is has not yet been the subject of standardization work. The service plane was not standardized for IN CS-1, i.e. no Q.1212 was produced.

1.2.1.2   The global functional plane

The global functional plane models an IN structured network as a single entity. It describes globally available service-independent building blocks (SIBs) (see Table 1.1). These SIBs are representations of the basic capabilities of the underlying functional network. There is one special SIB, the basic call process

(BCP), with which the other SIBs interact. Also defined are the points of interaction between the BCP SIB and the other SIBs. A set of global service logic defines, in terms of SIBs, a single service feature. The GSL describes where and how SIBs interact with the BCP.

Table 1.1   Service-independent building blocks (Q.1213).

Algorithm	Limit	Service data management	Verify
Charge	Log call information	Status notification	Basic call process (BCP)
Compare	Queue	Translate	
Distribution	Screen	User interaction	

1.2.1.3   The distributed functional plane

The distributed functional plane models a distributed view of an IN structured network. It is the first point where a traditional network architecture can be seen with the network being defined in terms of functional entities.

Each of the functional entities is able to perform actions. These are described as functional entity actions (FEAs). An FEA is the atomic unit of action and is wholly contained within a functional entity. FEAs are themselves comprised of functions (elementary functions (EFs)) which are the atomic units for FEAs.

Service-independent building blocks are realized within this plane as a sequence of FEAs. Each SIB is described by a set of distributed service logic which describes this sequence of FEAs. Ultimately some SIBs require informaton to be passed between functional entities to allow an SIB to be realized. It is these information flows that give rise to the protocol (intelligent network application part — INAP) in the physical plane (see Fig. 1.4).

1.2.1.4   The physical plane

The physical plane resolves the functional plane into a physical network and models the different physical entities and protocols that exist within an IN structured network.

For all the planes in the IN conceptual model there is a one-to-many relationship as you descend, i.e. a service comprises one or more service features; each service feature contains one or more SIBs as illustrated in Fig. 1.5.

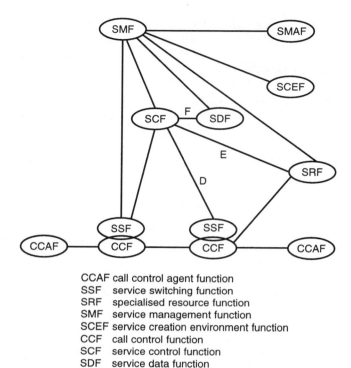

CCAF call control agent function
SSF   service switching function
SRF   specialised resource function
SMF   service management function
SCEF service creation environment function
CCF   call control function
SCF   service control function
SDF   service data function
SMAF service management access function

Fig 1.4   ITU-T defined functional IN architecture (distributed functional plane, based on Q.1211 [1]).

1.2.1.5   Mapping the elements of the INCM

Figure 1.5 shows how the various elements map in the IN conceptual model.

1.2.2   The IN functional architecture.

The distributed functional plane can be re-drawn to illustrate the IN functional architecture (Fig. 1.4), where the FEAs are grouped into functional areas, such as the service switching function, service control function, etc, which communicate through the use of information flows across reference points (reference points D, E and F are shown in Fig. 1.4).

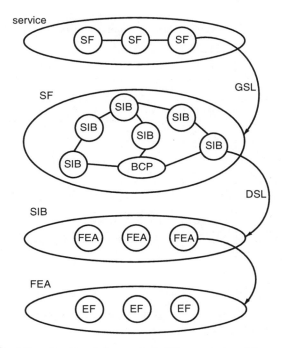

Fig. 1.5   Mapping of elements in the IN conceptual model.

ETSI, in 1993, produced 'Core-INAP' which is an operator-implementable version of the intelligent network application protocol for use in the D, E and F interfaces. The protocol is message-based and is defined in abstract syntax notation 1 (ASN.1) (see Chapter 3). The other interfaces shown in Fig. 1.4 have not been standardized in IN CS-1.

1.3   INTELLIGENT NETWORK COMPONENTS

Intelligent network functional entities, such as the service switching function, can be partitioned into a number of different physical elements, as shown in Table 1.2. This does not imply that other combinations of functional entities into a physical entity not shown should be disallowed.

The following sections provide descriptions of some of the elements identified in Table 1.2, such as the service switching point, service control point and service node.

Table 1.2   Mapping of functional entities to physical entities (Ref: Q.1205).

	SSF/CCF	*SCF*	*SDF*	*SRF*	*SMF*	*SCEF*	*SMAF*
SSP	Mandatory	Optional	Optional	Optional			
SCP		Mandatory	Optional				
SPD			Mandatory				
IP	Optional				Mandatory		
Adjunct		Mandatory	Mandatory				
Service Node	Mandatory	Mandatory	Mandatory	Mandatory			
SSCO	Mandatory	Mandatory	Mandatory	Optional			
SMP					Mandatory	Optional	Optional
SCEP						Mandatory	
SMAP							Mandatory

Figure 1.6 illustrates one possible IN configuration where each functional entity is mapped directly to a physical entity.

1.3.1   Service switching point

The service switching point (SSP) comprises the call control function (CCF) and the service switching function (SSF), as shown in Fig. 1.7. The CCF provides the call processing for basic telephony switching functions. The SSF, which is associated with the CCF, provides the interface to other physical entities, such as the service control point (SCP) and intelligent peripherals (IPs).

The SSP, as shown in Table 1.2, can optionally contain a service control function (SCF), specialized resource function (SRF) or service data function (SDF). If the SSP is the local exchange, it will also contain the call control agent function (CCAF). The separation of enhanced call control run on the SCP from service switching allows the progression of the call to be suspended while further information regarding how the call is to be handled can be obtained from the service data point (SDP). The decision to suspend call processing is based on the meeting of pre-specified criteria, i.e. triggers, for example dialled digits or line condition, at certain points during the call.

As shown in Fig. 1.6, SSP functionality can be provided at both local and trunk exchanges, the main differences being the selection of triggers available at each exchange.

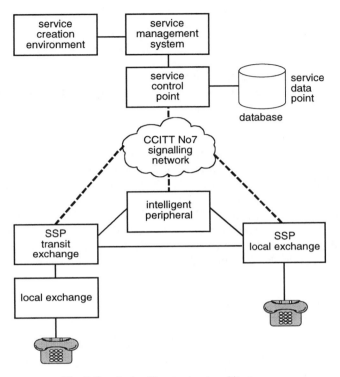

Fig. 1.6 An intelligent network architecture.

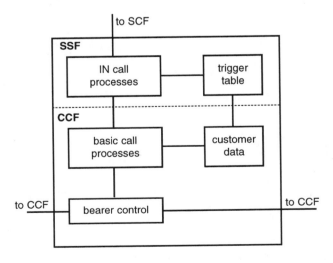

Fig. 1.7 Service switching point (SSP).

1.3.1.1   IN call model

In the separation of the call-control processing from the service switching functions in the SSP, the control system needs to have a reference model for the call from which enhanced service logic can be instigated, and to which it can also be returned, following the invocation of service logic. This model is referred to as the IN call model.

The IN call model is based on the definition of a basic call state model (BCSM), which defines the progression of the call in terms of points in call (PICs) and detection points (DPs). BCSMs are finite state models which reflect the originating and terminating sides of the call referred to as O_BCSM and T_BCSM respectively. The BCSMs as defined by ITU-T CS-1 are shown in Fig. 1.8, with the convention for illustrating BCSMs shown in Fig. 1.9. Call models also exist for the Bellcore-defined AIN. The differences between these call models add to the difficulty of interworking intelligent network equipment from different vendors .

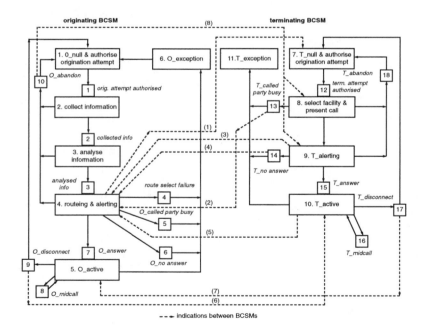

Fig. 1.8   ITU-T basic call state modles for originating and terminating half calls [1].

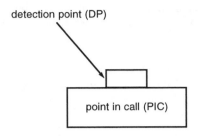

detection point (DP)

point in call (PIC)

Fig. 1.9   Point in call and detection point illustration.

1.3.2   Service control point

The service control point (SCP) is a general-purpose computing platform that hosts the advanced service control software or service logic.

The SCP should provide high throughput, high reliability/availability, fast response times and access to a high capacity database system. The SCP is usually a high throughput real-time computing platform on to which is loaded the SCP node software and a service logic execution environment (SLEE) (see Fig. 1.10) [2]. The SLEE shields the service logic from the underlying node/platform operations. The SCP node software provides the common utilities such as signalling system support, database communications, transaction monitoring and alarm reporting.

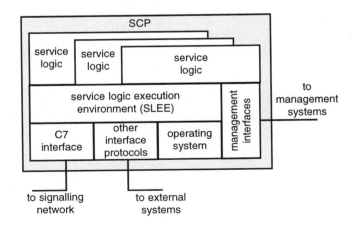

Fig. 1.10   Service control point.

Service logic is derived from the manipulation of a range of software building blocks (see section 1.2.1) within the service creation environment (SCE). It is the responsibility of the management system to deploy the resultant service logic to the network.

An SCP can be positioned in several locations in the network, for example connected to a number of SSPs via the signalling network, as in Fig. 1.6, or directly connected to the SSP as an adjunct processor via a high-speed interface.

1.3.3   Service node

The service node can provide intelligent network services, and engage in flexible information interactions with users. It communicates directly with one or more service switching points (SSP), each with a point-to-point signalling and voice circuit connection. Functionally, the service node contains a service control function (SCF), service switching point (SSF), service data function (SDF) and specialized resource function (SRF).

It provides a high availability, multipurpose platform of fully integrated resources (see Fig. 1.11), such as voice announcers, speech and DTMF recognizers, network-based answering, etc, that can be utilized during a call and shared across a number of SSPs in an intelligent network architecture. Under the control of service logic a caller would interact with the service node using each feature or service as required. The service node contains all the necessary logic and resources to deliver services to the customer.

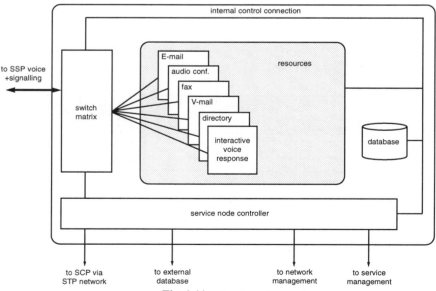

Fig. 1.11   Service node.

The service node provides an alternative service platform to the centralized SCP with network-based IPs.

The service node can be used as an 'edge-of-network' solution for services such as voice-mail. The system can also be used for number translation services where the service node originates the new call leaving the original switch in the speech path. The BT service node is described in detail in Chapter 6.

1.3.4   Intelligent peripherals

Intelligent peripherals (IPs) provide any additional specialized resource functions (SRFs) required in the network for the operation of services (see Fig. 1.12). Further details of speech systems as IPs are given in Chapter 5. The IP can also contain a service switching function/call control function (SSF/CCF) to provide switching to the appropriate resource in the IP.

Examples of IP resources include:

- customized and concatenated voice announcements;

- DTMF digit collection;

- synthesized voice/speech recognition;

- audio-conference bridge;

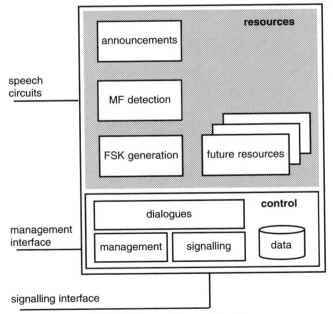

Fig. 1.12   Intelligent peripheral.

- information distribution bridge;

- tone generator;

- protocol converters;

- text-to-speech synthesis.

Depending on the actual network architecture, the IP can be either directly controlled by the SCP via the signalling network, with voice circuits connecting the IP to the SSP, or controlled from the SCP via the SSP.

The IP can be connected to more than one service switching point, allowing sharing of IP resources in the network — particularly important for expensive/specialized resources. The actual location of the specialized resources in the network depends on a number of factors, including cost, service operation, and performance.

1.3.5   Signalling network

The signalling network provides the communications between the component parts of the intelligent network, and is used to pass the controlling messages between the SCP, SSP, SDP (service data point) and IP. The IN application protocol (INAP) provides a standard message set for use in IN services, and is used for the interfaces between the SCP, SSP, IP and SDP. The protocol makes use of the non-circuit-related transaction capability provided by the transaction capabilities (TC) and signalling connection control part (SCCP) of the C7 signalling system (see Fig. 1.13). INAP is described in Abstract Syntax Notation No. 1 (ASN.1) in ITU-T Q.1218 and ETSI 'core-INAP' (see Chapter 3) [3, 4].

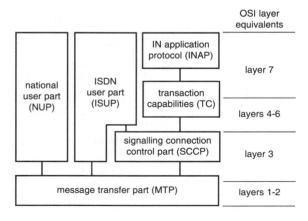

Fig. 1.13   CCITT signalling system No. 7.

1.3.6   Service creation environment

The ITU-T definition of IN CS-1 does not include standards for the service creation environment function (SCEF) in the same level of detail as for the service control point for example, but does identify the function and the links to the other functions, e.g. the service management function (SMF).

The service creation environment (SCE) provides tools for the rapid creation of new services using object-oriented programming techniques and reusable building blocks through the use of a graphical user interface (GUI). Also provided is a means to simulate the service before downloading to the network.

Through the use of SCE tools, the time taken to design a new service can be greatly reduced from that taken to implement new switch-based services, but in order to make this achievable the SCE tools must use building blocks which have already gone through rigorous testing.

Currently, service creation environments are closely coupled to the service control point with the SCP vendor providing the SCE which is matched to the architecture of their SCP.

A description of the processes and tools used in the field of service creation can be found in Chapter 8.

1.3.7   Service and network management systems

As with the service creation environment function, the service management functions are not standardized in ITU-T IN CS-1.

Currently, the standards bodies are investigating the harmony between functions in the intelligent network architecture and those defined in the telecommunications management network (TMN) standards.

The service management system, which is a combination of the service management function (SMF) and the service management agent function (SMAF), provides an operations support system responsible for service management, deployment, provisioning of customers and updating customer-specific data held on the SCP and service database. Additional functions which could be performed are database administration, network surveillance and testing, network traffic management and network data collection.

A description of service and network management functions related to intelligent networks can be found in Chapter 7.

1.4   INTELLIGENT NETWORK-SUPPORTED SERVICES

In the development of the IN CS-1 conceptual model, Q.1211, a number of example, or benchmark, services were used to identify the capabilities to be provided by IN CS-1. These services are often referred to as the IN CS-1 services, though they themselves are not standardized. They are only used for the validation of the concepts and should not be viewed as a limit to the services supported through an IN CS-1 architecture (see Tables 1.3 and 1.4).

The IN CS-1 benchmark services are in the category 'single ended', 'single point of control', and are referred to as Type A services. Currently, ITU-T is developing standards for Type B services, which will form part of IN CS-2.

Table 1.3   IN CS-1 benchmark services.

Abbreviated dialling	Destination call routeing	Selective call forward on busy/do not answer
Account card calling	Follow-me-diversion	Split charging
Automatic alternative billing	Freefone	Televoting
Call distribution	Malicious call identification	Terminal call screening
Call forwarding	Mass calling	Universal access number
Call rerouteing distribution	Originating call screening	Universal personal telecommunications
Completion of call to busy subscriber	Premium rate	User-defined routeing
Conference calling	Security screening	Virtual private network
Credit card calling		

There is a view that there are no real IN services, just services which are better supported in an IN manner. This can be as a result of the nature of the service or management of service data.

IN-supported services have one or more of the following characteristics:

- networked services — services that require co-ordinated action across a number of nodes;

- services that require large databases and real-time up-dates (especially by the customer);

- services that require tailoring to meet individual customer requirements;

- speculative services where the market demand is unclear and for which a trial platform is required.

Table 1.4   IN CS-1 service features.

Abbreviated dialling	Call queuing	Customized ringing
Attendant	Call transfer	One number
Authentication	Call waiting	Origin-dependent routeing
Authorization code	Closed user group	Originating call screening
Automatic call back	Consultation calling	Originating user prompter
Call distribution	Customer profile management	Personal numbering
Call forwarding	Customized recorded announcement	

One of the advantages of offering services in an IN manner is the reduction of the interaction between service features, which could cause unpredictable or confused operation. This is facilitated through the implementation of services on a common platform. Also, as the services are offered using a common platform feature, the interworking can be improved, leading to combinations of services or packages operating together, which reduce the need for multiple implementations of the same service for each platform in the network.

Some example IN services (described in the following sections) are:

- number translation;
- virtual private networks;
- cashless services;
- wide area centrex;
- automatic call distribution;
- personal numbering;
- call completion services.

1.4.1 Number translation services

One example of a number translation service is the provision of a single, fixed number for customers with geographically dispersed locations. Incoming calls

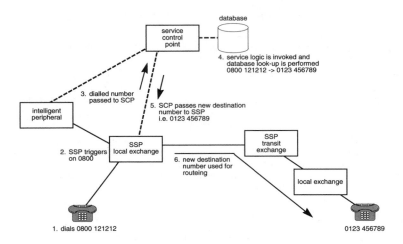

Fig. 1.14   Number translation service illustration.

will then be routed according to a predetermined schedule, taking into account, for example, nearest location, time of day or day of week.

Figure 1.14 illustrates an intelligent network implementation of a simple number translation service, where the SCP performs a database look-up to determine the appropriate destination number. The translated number is then sent to the SSP for the connection to the destination.

Number translation services are often linked to variable charging services, such as BT's FreeFone and Lo-Call services.

1.4.2   Virtual private networks

Virtual private networks (VPNs) allow users to create networks which can be considered private in terms of the flexibility, re-configuration, security, availability and exclusivity without having to manage dedicated private lines.

VPNs can be defined as a logical closed user group implemented on the public network, with the following main characteristics [5]:

- provision of a private numbering plan;

- call charging on the basis of duration usage as opposed to rental for private circuits;

- usage for switched voice, switched data or both.

Advantages of VPNs include a rich feature set (for example, call screening, authorization codes, flexible routeing), simplified network management tasks for users due to outsourcing, advanced billing facilities, and a single interface for domestic and international requirements.

VPNs can be accessed either directly through a dedicated link (on-net) or through a switched connection from the PSTN (off-net). Off-net access usually requires the use of special access numbers and authorization codes.

Virtual private network services include switched voice, switched data, private line, travellers or calling card access and 0800 access.

1.4.3   Cashless services

Cashless services, also known as alternative billing services (ABS) or calling card in the US, allow the customer to place calls which are then charged to different accounts. An example of this is the BT Chargecard service.

1.4.4   Wide area centrex

The term centrex is used to describe the provision of modern PABX services over the PSTN. Wide area centrex refers to the scenario where the customer has sites covered by two or more network nodes. A unified numbering plan and transparent operation of services are key features of wide area centrex.

1.4.5   Automatic call distribution

Automatic call distribution provides the distribution of incoming calls to a number of separate answering points, these may also be geographically separate, either on a fixed proportional basis or dependent on availability of answering points. This service is widely use by ticket booking agencies, operator positions and airlines, and is often associated with a single directory number.

1.4.6   Personal numbering

This refers to the provision of a single number for the customer, with the network providing the capability of call completion based on an updatable personal profile held in the network. The call could be completed to the person's actual or most likely location with a voice mailbox or paging system being the default option if the person is busy.

1.4.7   Call completion services

These are simple diversion services, such as divert-on-busy or divert-on-no-reply, and those already offerred as switch-based services, for instance BT's network services.

1.5   EXAMPLE OF AN INTELLIGENT NETWORK IMPLEMENTATION

Many telecommunications operators have installed or are planning to install intelligent network equipment to meet the needs for specific service requirements, such as Freefone, premium rate services and calling card services.

An example of an early IN implementation is the BT digital-derived services network (DDSN) which provides the advanced number translation services based on AT&T IN technology and installed in 1989. The DDSN consists of an overlay network of 13 AT&T 5ESS switches acting as SSPs and three pairs of network control points (NCPs) providing the service control function.

The advanced number translation service includes the following features:

- time-of-day/day-of-week routeing;

- geographical routeing — routeing depending on area where call originates;

- proportional routeing — based on a predetermined routeing plan;

- call queuing/completion of busy calls;

- call prompting — uses voice guidance and digit collection.

In 1992, IN Phase 1, an SCP consisting of a GPT System X front end and a Tandem computer, was introduced into the BT network. The IN Phase 1 platform is used to provide basic number translation services.

1.6   THE FUTURE INTELLIGENT NETWORK?

In the standards arena, the harmonizing of the developments of AIN, ETSI and ITU-T standards will begin to occur. This may ease the possible problems when interworking equipment in a multivendor environment.

Significant developments are occurring in the field of speech recognition, which, linked with IN capabilities, will move services offered on the network towards the goal of a network to which you can talk. Speech recognition could be initially used in the provision of a voice dialling system, leading on to more inter-active-based services, making use of word-spotting capabilities of the systems.

Looking to the future, it should be noted that some people do not consider the standards-based IN described here as the intelligent network in the sense that it has not yet recreated the user-friendly interface of the operator-based networks with a network to which you talk and which talks back, offering you what you want when you want it [6].

Intelligence can exist in a number of locations in the network; outlined so far has been the use of a centralized IN architecture, but this excludes the intelligence found in modern customer premises equipment (CPE), such as smart-phones, PBXs and computer telephony integration (CTI), etc. As intelligence in the network becomes more complex, intelligent CPE will be required to make use of some of the enhanced features. A key issue with the distribution of intelligence is that of feature interworking and interaction (desirable and undesirable respectively) in a distributed processing environment. Chapters 4, 10 and 13 discuss these areas in more detail.

IN concepts, though currently applied to telephony-based networks, are also applicable to other traffic types, such as data, broadband, video and multimedia services. Over the next few years this area will see significant attention to development and growth.

With the transformation of the telephony network to a 'data' network and the increase in information networking, the form of future intelligent networks will be different from those of today, implemented through a process of both evolution from the existing installed platforms, and revolution via the integration of new technologies and concepts.

REFERENCES

1.   ITU-T Recommendations are available from: ITU Sales Section, Place des Nations, CH-1211 Geneva 20, Switzerland (Tel: +41 22 730 51ll, Fax: +4122 730 5194).

2.   Eburne M: 'Intelligent networks', IBTE Structured Information Programme (1992).

3.   Interface Recommendation for Intelligent Networks Capability Set 1, Q.1218, ITU-T (June 1992).

4.   Core Intelligent Network Application Protocol (INAP), Version 8, ETSI (May 1993).

5.   'Intelligent networks: strategies for customised global services', OVUM (1993).

6.   Dufour I G: 'Coping with incomplete standards', Intelligent Networks Industry Forum (September 1993).

2

INTELLIGENT MOBILE NETWORKS

N C Lobley

2.1   INTRODUCTION

The past few decades have seen a massive increase in the use of mobile communications. Cellular telephones are commonplace within society and can be seen in use in all areas of modern life — at home, work and in public. The true dawn of mass-market mobile communications is only around the corner. By the year 2000 it is expected that the small hand-held pocket communicator will be in use by a significant proportion of the population.

Mobile communications has been made possible by advances that have enabled compact radio technology and computing intelligence to be added to user's terminals and telecommunications networks. Telecommunications switches have evolved to include computer databases and complex interfaces to handle interactions with radio-equipped base-stations. Functionality has been added to enable the networks to track, locate and provide continuous communications for the users while they are mobile. Added intelligence in the terminal has enabled the traditional wire-line link to the switch to be replaced via radio. Intelligent schemes for optimizing the use of the precious radio resources has also enabled high user-traffic densities to be achieved.

2.2   ANALOGUE SYSTEMS AND THE CORDLESS TELEPHONE

The origin of land mobile communications appeared around 30—40 years ago with vehicle-mounted radio transmitters. Mobile communications have made massive leaps forward since then, with developments being split between two basic areas:

- 'cellular', where a telecommunications network is used to enable continuous communications over a wide area;

- 'cordless', where a more concentrated local coverage is provided over small 'pockets' — cordless is generally provided independent of the fixed network [1].

2.2.1   Analogue cellular

The contemporarily established first generation mobile systems within the majority of the world are based upon analogue cellular technology, which is now starting to roll over towards digital technology. The common systems include the North American based AMPS (Advanced Mobile Phone Service), the Scandinavian NMT (Nordic Mobile Telephone) and TACS (Total Access Communications System) within the UK, which is a variant of the American AMPS. Analogue speech transmission with frequency modulation operating in the 450 or 900 MHz bands is used for these analogue systems with the network components based around proprietary manufacturer's nor national standards. With no internationally agreed network nor radio interface standard, network vendors and operators had little flexibility in the production and procurement of network components. This problem has been resolved by the development of second-generation digital systems. Within Europe the second-generation standard for cellular communications is called GSM — the global system for mobile communications [2]; within the USA, digital developments of AMPS have produced the American IS-54 and IS-95 second generation systems.

2.2.2   Cordless communications

Within the cordless area, developments were driven following the development of simple analogue cordless telephones. The early 1980s saw the potential flooding of cheap imported units to a wide variety of frequency standards which offered tetherless communications from the handset to the base unit. Moves were made within Europe to develop standardized cordless communications. The initial development using digital communications was CT2 (Cordless Telephone 2), and offered improved quality and capacity on the radio path with a common air interface enabling handsets to 'roam' between cordless systems. The DECT (digital European cordless telecommunications) system included the support of ISDN and LAN data capabilities in excess of 384 kbit/s full duplex. CT2 operates around 900 MHz while the DECT standard was developed to operate at around 1.8 GHz. CT2 and DECT standards do not define the network aspects for the system; however, DECT does include the capability to support wide area mobility, if linked by a mobility-supporting network such as GSM.

2.3   CELLULAR SYSTEM ASPECTS

The term 'cellular' is derived from the realization of the system. Cellular comprises a number of transceivers geographically dispersed to give separate coverage areas or 'cells'. The type of layout produced optimizes the radio spectrum resources by enabling radio frequencies to be re-used in non-adjacent cells. Terminal mobility is the feature of cellular systems that is the major difference from fixed telephony. Cellular systems offer mobile users the capability to make and receive calls from their mobile telephone while within coverage of the network. The cellular system automatically tracks and records the location of the terminal within the system to enable mobile terminating calls to be delivered. Once calls are in progress between the terminal and the network, communications can continue while the mobile terminal is moving geographically. This can mean the terminal crossing radio coverage boundaries (moving between cells) and requiring a 'handover' to take place.

Mobility places major requirements upon the mobile network. The feature enabling the system to track the location of the subscribers is termed 'mobility management'. The link between the terminal and the network is subject to constant changes due to the requirements of the radio link. The quality of this radio link can change due to interference, path changes and system operations. The flexible management of the radio aspects to cater for this is termed 'radio resource management'.

To enable calls to be routed to the subscriber, the system must be able to locate the subscriber's terminal. The terminal will automatically inform the network of its current location via the 'location update' procedure as it moves around within the radio coverage offered by the system. The network groups cells into location areas, and the cells within the location areas broadcast location area identity information which the terminal then returns to the network, along with its own identity. From this information the network database can update the database entry of the subscriber with the current location. For a mobile terminating call, the dialled digits of the called subscriber will be used by the network to query the network database. The network will obtain the current location information held within the database and use this to initiate the paging procedure. The network will broadcast the subscriber's identity throughout the current location area. For a successful paging attempt the terminal will respond, and communications between the terminal and the network can be initiated. The network then uses this point of communication to deliver the call via internal routeing.

In the case of a mobile originated call, the subscriber's terminal will attempt to make a call. The network will check the subscriber's database information to ensure the call may be made. Then, subject to any restrictions, the call will be allowed to continue as normal.

Due to the mobile nature of the subscriber, terminals may move during communication, possibly requiring the serving cell to change. To cater for this the handover procedure (also termed 'hand-off') is used to enable the call to pass between cells without 'dropping out'. The handover may require a change of both radio path and network path which should not noticeably interrupt communications.

Cellular systems place major requirements upon the networks which require added intelligence and processing capabilities to manage handover, mobility management, and the subscriber's call-handling profile details.

2.4   THE GLOBAL SYSTEM FOR MOBILE COMMUNICATIONS

As introduced earlier, nearly all of the countries within Europe currently offer some form of mobile telecommunications. The demand for a European-wide standard mobile communications system was created by the limitations of operators' early analogue systems. National solutions had led to expensive non-standard network and terminal equipment. The diversity of these national standards and the limitations on users being able to roam within mainland Europe (to make and receive calls in different countries), in conjunction with varying quality, coverage and capabilities, encouraged the start of developments of this system in 1982.

The result was GSM, the 'global system for mobile communications', which was initially developed within the 900 MHz band; however, the availability within the UK of spectrum at 1800 MHz enabled the scope of GSM to be extended to cover this band. The developments mainly revised the radio aspects, with the system being referred to as DCS 1800 (digital cellular system 1800). In this form, the standard has also been applied within Germany and Hong Kong, and also been revised for 1900 MHz application in the USA. The American derivative is being rolled out as a TDMA personal communications services solution in parallel to other CDMA-based developments. It is worth noting that some USA CDMA operators have chosen the GSM network architecture to support the CDMA radio interface.

2.4.1   GSM services and features

The primary role of GSM was to provide mobile voice telephony. However, the system offers the capability of data services to complement ISDN at rates of up to 9600 bit/s full duplex. Other capabilities include the short message service (text messages), facsimile transmission and packet data access. The roaming capability throughout Europe is automatic — upon switching on the GSM phone in any area covered by GSM, the 'visited' network will notify the 'home' network of the location and obtain the subscriber's parameters. Thus mobile originating and mobile terminating calls can be made.

Many of the early analogue systems encountered security problems with cloning of terminals and terminal theft. The GSM system has made significant improvements in the security area. The GSM subscription details are contained on a smartcard called the SIM (subscriber identity module); this has to be plugged or installed into the terminal before use. The GSM system provides digital encryption and authentication of communications, and can also check for type approval of terminals used on the network. Handset size developments have been a major factor in the success of GSM. Approximately 8 to 10 years were needed for analogue handsets to evolve from large 'transportable terminals' to pocketable phones. GSM terminals are already available in pocketable form despite being relatively new within the market-place. Figure 2.1 shows the progress made with GSM terminals. A GSM 'SIM' card is also shown.

Fig. 2.1   GSM SIM card and terminals.

2.4.2   The GSM standard

As a standardized system GSM not only defined a common radio interface but also a network architecture which enabled network vendors to develop equipment to conform to the common standard, and also enabled network operators to purchase equipment for use within the network from a variety of vendors. These network components should all interwork correctly due to conformance with the standard.

2.4.3   The GSM network architecture

The GSM network has been one of the major successes of European standardization. Its standardized architecture has allowed a variety of options for manufacturers and operators to introduce GSM. The GSM architecture is shown in Fig. 2.2. In many respects, the GSM architecture is a good example of an early intelligent network (IN) architecture as it follows the concept of separating switching and service-related functions (see Chapter 1).

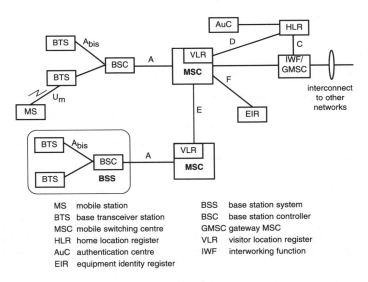

Fig. 2.2   The GSM architecture.

Procedures for call handling, mobility management, handover, roaming and location management have all been defined using the architecture. The functionality of the various network entities is described below.

The **mobile station** (MS) consists of the physical equipment used by the GSM subscriber. The MS contains the subscriber identity module (SIM), which stores information such as the subscriber's addressing information, authentication parameters and security functionality.

The **air-interface** (Um) is the interface between the MS and the BSS.

The **base station system** (BSS) is the physical equipment providing the radio coverage. Functionally, the BSS is divided into two parts:

- the **base station controller** (BSC) which provides the control function and interfaces with the mobile switching centre (MSC) via the A interface, and is responsible for identifying the need for handovers and managing intra-BSC handovers;

- the **base transceiver station** (BTS) which provides the radio transmission functions for the radio coverage in the cell.

The **A-bis interface** between the two entities has been defined, enabling the two functions to be implemented separately, if required.

The **A interface** is defined between the BSS and the MSC and carries information for the base station system application part (BSSAP) protocols. The BSSAP consists of two separate applications:

- the base station system management application part (BSSMAP) protocol which is used to support processes in the MSC and BSC;

- direct transfer application part (DTAP) which is used to relay messages between the MS and the MSC transparently.

The **mobile switching centre** (MSC) carries out the switching functions needed for all mobile stations within a geographical area controlled by the MSC. The MSC takes into account the mobile nature of the subscribers by managing the radio resources during the inter-BSC or inter-MSC handover process. It is also involved in the location update procedure, when the MS moves into a new area.

MSCs can have an interworking function (IWF) associated with them for interconnection purposes. The functionality of the IWF depends on the type of network to which the interconnection is required and also the services requested. Specially designated gateway MSCs (GMSCs) have interrogating node functionality which is used to access the HLR for routeing and service-related informaion for mobile terminating calls.

The C, D, E, and F network interfaces within the GSM network, shown in Fig. 2.2, are based on the CCITT Signalling System No. 7 (SS 7) and use the mobile application part (MAP) signalling procedures.

The interfaces between MSCs and the interconnection between the GSM network and other networks use standard connection-oriented protocols such as ISUP, TUP/NUP for call-handling purposes.

The **visitor location register** (VLR) is the database containing information needed to handle the calls set up or received by the MSs registered with the VLR. Within early versions of the GSM standard the 'B' interface was used between the MSC and the VLR; this interface has now been subsumed, with the VLR being integrated into the visited MSC. A mobile station roaming in an MSC area has its current location data managed by the visitor location register. The VLR is in charge of the data records of all the MSs registered within its catchment area. Certain requests involving the supplementary services that require additional information are met by the interrogation of the home location register.

The **home location register** (HLR) is the database responsible for the management of the mobile subscriber's records. It manages information related to service subscription profiles, e.g. teleservices, bearer services and the supplementary services that are to be made available to the MS. It also dynamically stores the current location information such as the visited MSC/VLR of roaming mobile stations, allowing the appropriate routeing of mobile terminating calls.

The **authentication centre** (AuC) is a database concerned with the security aspects. It is responsible for managing the allocation of individual MS's authentication parameters from the home network to the visited network. These are used by the visited network for authentication of the visiting MS, and establishing ciphering on the radio channel.

The **equipment identity register** (EIR) is a database which stores equipment identities which are used by the GSM system to identify unique pieces of mobile equipment. In general, the EIR is used to check if mobile terminals have been stolen.

2.5   FUTURE MOBILE SYSTEMS

GSM has been a resounding success within Europe and the world. By offering a common standard for both air and network interfaces, operators have been able to provide digital cellular systems with extensive service capabilities. Despite the success of second-generation systems, many limitations still exist. Developments are being made to define the future third-generation systems which will cater for the mass-market mobile communications needs beyond the year 2000. Within Europe, this system is called universal mobile telecommunications system (UMTS), globally being referred to by the ITU as the future public land mobile telecommunications system (FPLMTS).

The major aim of UMTS is to define an integrated system capable of providing a variety of mobile services (with a data rate of up to 2 Mbit/s) to encompass those currently offered by (dedicated) cordless, cellular and satellite mobile systems. UMTS will offer mass-market mobile communications to its users enabling access to services in the domestic, business and public (cellular) environments by using the same small 'pocketable' terminal.

2.5.1   Network requirements for UMTS

Mass-market fixed communications are currently provided by the PSTN GSM, and first-generation analogue systems offer cellular services, with cordless being widely used for tetherless access. The future telecommunications user will wish to access communications services from a common terminal. UMTS will be the step forward that enables this.

From the network evolution point of view, fixed networks are moving towards intelligent network (IN) based N-ISDN (narrowband ISDN) and B-ISDN (broadband ISDN) operation. Operators are adding intelligence into the network using INAP (intelligent network application part of SS 7 — see Chapter 3) and SCPs (service control points). IN service creation (see Chapter 8) and delivery techniques are also developing to enable network operators to offer a large variety of customized services.

The GSM mobile network is enabling digital mobile capabilities to be provided which are feature-rich in functionality. At present GSM mobile networks are providing services which are not widely available within fixed systems. As a stand-alone network GSM relies upon mobile-specific network components (MSC/VLR and HLR). The same switching infrastructure cannot effectively be used for both fixed and mobile at present. The GSM mobility management and handover functions have been integrated into the switch (MSC). This has substantially increased some of the processing requirements to the switch, which in realistic terms has severely reduced the subscriber capability of GSM MSCs compared to equivalent fixed switches.

Due to the higher frequency of operation and large capacity requirement of UMTS, a dense cell architecture is required to provide adequate mass coverage; this will create large network infrastructure costs if the service, mobility management and handover features are developed around switch platforms in the same way as GSM. From the operators and equipment vendors perspective, the stand-alone network approach is not feasible. The common aim for UMTS is harmonization of the system's components via integration of mobile and fixed capabilities into a common architecture. This architecture consists of the core backbone switching network providing mobile call handling, with handover and mobility management being catered for by added intelligence — the intelligent network.

2.5.2   The IN solution for UMTS

The IN concept plays an important role in existing and future telecommunications systems. It is capable of providing simple, supplementary services that can be extended further to enable the implementation of advanced services and facilities in a flexible and efficient way. IN is seen as the solution for UMTS to enable network operators and equipment suppliers to produce integrated platforms which will cater for fixed narrowband and broadband communications and include mobile capabilities for UMTS. This minimizes network infrastructure costs by reducing the number of network platforms within an operator's architecture while enabling common service provision for all end users, whether fixed or mobile. The IN approach will enable evolution of the GSM architecture to include the IN flexible service creation and delivery capabilities. A major aim of this approach is to enable UMTS to be added to networks via the addition of advanced intelligent capabilities to both the switching and the radio access components.

2.5.3   IN capability set 1

Within fixed networks, IN is being applied to provide a variety of functions, mainly related to the provision of simple telephony-based supplementary services. The IN capability set 1 (CS-1) standards use the service switching functions (SSF) and call control functions (CCF) to interact with the service control function (SCF) (which is supported by other functionality) to provide services. The CS-1 SCF is a single point of control and relies upon the SSF/CCF combination to invoke action. If added capabilities such as DTMF receivers, conference bridges or voice announcements (specialized resources) are needed, the SCF can switch in the specialized resource function (SRF) to provide these features. Although not fully achieved, CS-1 could be viewed as an architectural concept capable of providing additional capabilities to facilitate advanced service provision from a network-independent platform. To provide the advanced terminal mobility which is essential for UMTS, a variety of mobile-specific requirements must be satisfied.

2.5.4   UMTS procedures

As a mobile system, UMTS utilizes many fixed and mobile procedures. Studies of these procedures have identified a variety of requirements to the network operation.

2.5.4.1   Call handling

For mobile originating calls, the subscriber's service profile must be available within the local network (where the subscriber has roamed) for the visited network to verify the subscriber's requests for services and features. This will require inter-network communications to enable the profile transmission between the home and visited network.

Call handling for mobile terminating UMTS calls will use the separation of call and connection control. This is a concept introduced within broadband communications developments which is very applicable to mobile communications where the location and status of the called terminal may not be readily available when the call is originated. For UMTS-terminated calls, the call processing should be halted at the earliest possible time and the locating procedure (to find the position of the called terminal) commenced. This will involve the switch-passing details of the called (UMTS terminal) number and the requested service to the SCF. The SCF will recognize that the dialled number is for a call destined to a UMTS number and will initiate locating by querying the called profile within the database (SDF). The profile holds details of the current location, recorded status (busy, free) and details of the terminal and subscription capabilities. This enables the call to be failed, forwarded or routed at the earliest opportunity with the optimum use of the network resources. For a successful call, the SCF will communicate with the distant (visited network's) SCF and initiate paging of the called terminal. The outcome of the paging will be a network address for the called terminal which will be communicated back to the calling SCF. The calling SCF will instruct its local switch to route to the called terminal and call processing can continue.

2.5.4.2   Mobility management

Within IN CS-1, all procedures are call related, relying upon a state machine (the basic call state machine) within the switch to 'trigger' and invoke IN processing. UMTS mobility management procedures such as location registration, location updates and paging of called terminals can occur outside an active call; consequently, alternative (non-call related) techniques to invoke the IN processing are required. The intelligent network must provide the control of these actions and enable the UMTS terminals to interact with the SCF entities directly. Obviously, the terminal to SCF communications must be secure to maintain network integrity. This direct application of terminal to SCF communication illustrates the far-reaching enhancements to IN that are needed for UMTS.

2.5.4.3   Handover

Handover is the terminal mobility procedure which is likely to place the most substantial impact upon the network aspects for UMTS. Mobile terminals send and receive traffic to and from a particular base-station connected to the network when calls or signalling associations are in progress. Within contemporary systems handover is only used to satisfy speech and data quality requirements due to a changing radio path. Within UMTS a handover may also be initiated at the request of the user (e.g. for subscription-related reasons) or the network operator (typically for network management purposes), irrespective of link quality.

Rerouteing a call to a new base-station is a significant task within the fixed network. Figure 2.3 shows a typical tree-like network topology supporting UMTS, in which the base-station sub-system (BSS) comprises an arrangement of base-stations and associated switching and control functionality. Often, handovers will take place between base-stations of the same BSS, but they may also occur between base-stations belonging to different BSSs, local exchanges (LE), or even transit exchanges (TX). During the process of handover, a common point has to be determined for bridging of the bearers between the old links and the new links, in order that the mobile terminal can make a seamless (unnoticeable) handover from the old to the new. As a consequence of the tree-like structure of the network, this bridging point may be located at any level in the network. Thus handover has a large impact on the fixed network [3]. These impacts may be minimized if specific functionalities are split between the fixed network and control components. Use of IN control principles for handover within the core network will remove the processing burden of handover control integrated into the switches.

It is particularly important that handover is achieved rapidly (before a degrading radio channel is lost completely), and that the signalling load imposed on the network to provide handover is minimized.

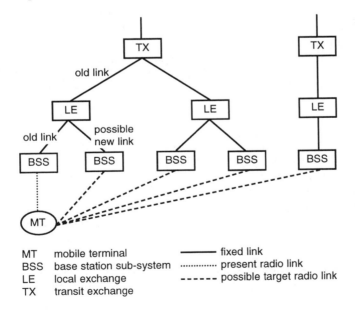

Fig. 2.3   Handover possibilities within the network.

2.5.5   Developments of IN for UMTS

To satisfy the needs of UMTS, and handover in particular, IN will have to evolve significantly from the current CS-1 position [4]. To provide UMTS, several major enhancements are required. The major developments are introduced below.

- Support of non-call-related functions — certain mobility-related functions (such as location management and handover) will require automatic invocation by the terminal when no call is in progress or without user intervention.

- Feature repeatability — the mobility features may need to be repeated several times during a call, e.g. a handover may be performed (at different levels of the network hierarchy) more than once.

- Concurrent and distributed execution of IN features — the UMTS terminal may need a number of simultaneous IN features (e.g. handover and location updating) to be performed. IN should be enhanced to support the execution of a variety of independent and concurrent IN-based services. The execution may occur at differing points within the network.

- Transfer of control — a UMTS terminal may move between the coverage areas of control entities during the execution of a service; this may require the service logic to be transferred mid-execution, or halted, transferred to another point of control and restarted. A typical example of this may be the handover procedure which will require interaction and co-operation between SCFs. If the handover of a call occurs between switches, the active BCSM (call state machine) will need to be passed from the old to the new switch. This should include conveyance of any active or armed trigger detection points.

- Advanced service interaction management capabilities — there are a number of areas where mobile features (such as handover) will have an impact on the provision of IN services. The contention between the handover procedure and call-waiting supplementary service is one example of interaction that requires management. If the local exchange involved in the call must change due to handover, it is essential that the IN service logic and switch-based detection point triggers catering for call waiting are also moved between switches during the handover. The interaction between SIBs responsible for handover and other IN services is more easily managed than an interaction between IN-based services and non-IN services.

- Intelligent terminal equipment — it must be possible to provide limited IN functionality in the mobile terminal, in order to support intelligent execution of handover and other UMTS services and procedures.

To achieve the full benefits of an IN approach to UMTS the evolving IN architecture requires distribution of intelligence within the core network, access network (base-station system) and mobile terminal equipment.

2.5.6   UMTS functional architecture

Figure 2.4 shows the IN-based functional architecture for UMTS [5]. The structure of the functional architecture for UMTS, including the allocation of functionality within it, is an essential step in defining the respective roles of IN service control and the supporting fixed network.

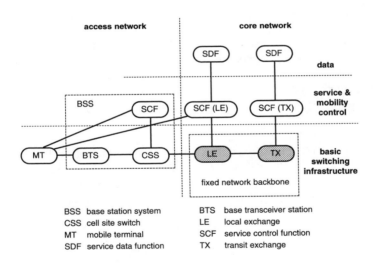

Fig. 2.4   An IN-based UMTS functional architecture.

An important characteristic, which structures the architecture in line with IN concepts, is the separation of the UMTS service and mobility control, and data from the basic backbone switching infrastructure. The backbone network is represented by the LE and TX switching functional groups. These will meet the needs of both fixed and mobile communications services, but are not dedicated to UMTS alone.

A basic guideline, to ensure a generic backbone network, is that any UMTS-specific functionality should be removed from the LE/TX functional groups, as these represent basic switching capability. The traditional way of dealing with mobility procedures such as handover is to provide special mobility functionality in the switches. This, of course, ties the mobility functions to one particular network, thereby limiting scope for implementation, and results in the switches becoming UMTS-specific, rather than general purpose. For UMTS, the handover control functionality is therefore removed from the switching infrastructure and placed in the service control functions where it can be provided as an IN application.

The structure of the functional architecture is also influenced to a large extent by the requirements of the UMTS mobility procedures, and handover in particular. The main requirement of distributed control is reflected in the distributed nature of the network. SCFs at the BSS level as well as at the LE and TX level allow handover control functionality to be as close as possible to the bridging point, with the consequence that signalling links are as short as possible. The SCF has been distributed to cover the base-station system, local exchange and transit exchange. This is essential in UMTS, where the time-critical nature of some operations, such as handover, requires that functions are performed close to the point of invocation. The roles performed by the individual SCFs may be very dependent upon their location, the core SCFs (relating to the switching) will include functionality very similar to contemporary IN CS-1 SCFs. The SCF CSS and the CSS within the BSS will fulfil the role similar to the base-station controller within GSM of providing intra-BSS control and switching for handover purposes.

The SCFs found in the architecture will have the ability to process functions in a manner that will allow continuity of SIB chains across physical areas. This is illustrated in Fig. 2.5 which shows the distribution of handover control functionality over different control functions within the network. The handover control can then be transferred between SCFs as the bridging point changes.

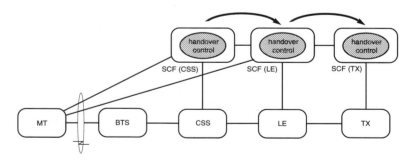

Fig. 2.5   Transfer of handover control within the network.

The functional architecture includes the mobile terminal (MT), which in UMTS will require a degree of IN functionality to enable support of advanced services such as handover. The MT represents one in a class of intelligent terminals which will carry limited service control functionality in order to support the advanced services of the future.

2.6 CONCLUSIONS

Mobile system developments to date have been reviewed, with emphasis placed upon the network components. The GSM architecture has been discussed, with the future network requirements and developments for UMTS being outlined.

The current stand-alone mobile and fixed architectures will be unable to cater for the major impacts that the future mass-market mobile communications system, UMTS, will require. Significant enhancements are required to intelligent, fixed and mobile networks to provide the integrated communications network of the future. The intelligent network developments should include:

- non-call related processing;

- concurrent and distributed intelligence with transfer of control;

- the addition of more intelligent terminals to satisfy UMTS requirements.

Only if the UMTS requirements are included in the intelligent networks of the future will the significant burden of mobility management and handover be achievable. These can be provided within the same core switching architecture that can offer flexible service development and creation while minimizing the major problem of services and features interaction.

With the inclusion of the network developments outlined, UMTS will be well placed to provide service to users (who are mobile and stationary) well into the twenty-first century. This will have been achieved in the most efficient and cost-effective manner to both equipment vendors and network operators.

REFERENCES

1. Mouly M and Pautet M: 'The GSM System for Mobile Communication', ISBN 2-9507190-0-7 (1992).

2. GSM standards, ETSI, Sophia Antipolis, France.

3.   Mason P C, Brydon A N and Cullen J M: 'UMTS handover requirements in the context of an intelligent network architecture', Fourth IEEE International Symposium on Personal, Indoor and Mobile Radio Communications, Yokohama, Japan (September 1993).

4.   ITU-T Q.1200 series IN CS-1 Standard (1993).

5.   R2066 deliverable: 'UMTS Network Architecture Draft', R2066/FACE/GA3/DS/ PO33/b1, Commission of the European Community (June 1992).

3

SIGNALLING IN THE INTELLIGENT NETWORK

M C Bale

3.1   INTRODUCTION

To achieve the goal of rapid service creation and deployment in public switched networks, it is necessary to functionally separate the control of services from basic call processing. The intelligent network (IN) supports this concept, and defines an architecture of modular and reusable network functions that enable service/network independence. Physical distribution of these functions is also supported, giving a network operator the flexibility to build a network using platforms that are specialized to perform specific network functions.

Distribution brings with it the need to communicate. It is essential that the protocols used for signalling between the network functions are service independent, since this allows new services and features to be implemented without upgrades to the signalling systems. Standardization of the signalling systems is also important, particularly in a multi-vendor network or where interconnect to other networks is required.

In building public telecommunications networks, network operators have developed digital signalling systems that are efficient and resilient. However, many protocols have been shown to be somewhat inflexible, monolithic and specific to particular operators. At the same time, the computer industry has pursued 'open standards' and 'open systems'. Open standards and open systems (systems which conform to the ISO open systems interconnection reference model — OSI) allow distributed processes to communicate and co-operate across one or more networks. Furthermore, they propose modular and layered approaches, which give the adaptability required in a modern signalling network.

The last few years have seen the maturity of the OSI model, as well as significant developments in computer and software technology. The need for flexibility in network signalling systems has led to the adoption of OSI principles, reflected in the current international standards for Signalling System No 7 (SS7) and the intelligent network application protocol (INAP).

3.2   SIGNALLING SYSTEMS

The software processes in today's integrated digital networks communicate to control the network using common channel signalling system No 7 [1-3]. This is a digital, message-based common channel signalling system, primarily for use within the network.

The fact that SS7 is digital and message-based makes it ideal for inter-process communication. Being common channel, it is logically (and can therefore be physically) separated from the bearer circuits to which it relates. This gives a greater degree of flexibility in the signalling network, allowing high functionality, resilience and performance.

Signalling between the network elements in an intelligent network requires the versatility of data communications protocols as well as real time efficiency. These factors have been the main drive behind the more recent developments of SS7, particularly the need to support:

- circuit-related signalling — the transfer of information which relates to the establishment, supervision and release of bearer circuits;

- non-circuit-related signalling — the transfer of information which does not relate to the switching of connections; this information transfer is required to provide services (e.g. for communications between the call-control processes and the processes controlling the supplementary services), and is primarily transaction based.

Recommendations for SS7 are produced by the telecommunications standardization sector of the ITU (referred to as ITU-T, formerly CCITT), and are standardized by regional standardization bodies, such as the European Telecommunications Standards Institute (ETSI).

Currently, the majority of network users have analogue access to the network. Signalling between the user's equipment (e.g. telephone) and the network is based on on-hook/off-hook detection at the local exchange, and pulsed or DTMF (dual tone multifrequency) dialling within the speech band (bearer circuit). In many cases, digital access is also supported, usually as an ISDN user-network interface (UNI). This supports common channel, digital, message-based signalling, referred to as Digital Subscriber Signalling System No 1 (DSS1) [4]. Recommendations for DSS1 are again produced by the ITU-T and standardized by the regional standardization bodies.

3.3   PROTOCOL REFERENCE MODELS

The evolution of the ITU-T digital signalling systems has followed the development of the ISO open systems interconnection (OSI) reference model [5]. The OSI protocol reference model defines a seven-layer logical structure for the development of protocols used for communications between open systems. For each layer in the model, the purpose of the layer, the services it offers, and the services it requires from the layer below are described. Routeing and addressing within the structure are also described.

The specification of SS7 predates the OSI model. While it does have a layered structure, it consists of only four levels. However, an important aspect of the SS7 protocol reference model (shown in Fig. 3.1) is that there is a clear separation of the application process (user part) from the underlying protocols (message transfer part — MTP — which is common to all of the user parts). The function of the message transfer part is to provide a reliable signalling connection between any two points in the network. To do this, it comprises three levels:

- signalling data link (MTP level 1) — provides a physical bi-directional transmission path between two directly connected network elements;

- signalling link (MTP level 2) — provides a reliable logical signalling channel between two directly connected network elements;

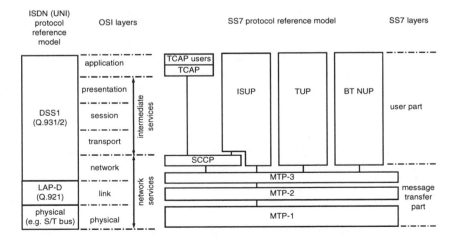

Fig. 3.1   SS7 and ISDN (UNI) protocol reference models.

- signalling network (MTP level 3) — provides routeing of signalling messages from any point in the SS7 signalling network to any other point; the routeing of messages is based on the MTP point code, which uniquely identifies an SS7 signalling terminal in the network, and the service information octet, which identifies the user part, or MTP level 3 at that signalling terminal (the MTP level 3, also provides for traffic, route and link management of the signalling network, for example automatically re-routeing a signalling connection on detection failure at level 2).

A number of SS7 user parts have been defined by ITU-T. These include the telephony user part (TUP), to support basic telephony, and the ISDN user part (ISUP), to support ISDN call and connection control. To support early implementation of SS7 in the BT Network, BT developed an enhanced version of the TUP, known as the BT national user part (BT NUP). This user part supports basic telephony and some ISDN features. Each of these user parts is specified as a single layer structure.

The signalling connection control part (SCCP) was added by ITU-T as a user part to the SS7 protocol reference model to align SS7 with the OSI model. The SCCP provides an OSI network layer conformant interface to the MTP, enabling OSI conformant higher layers to be used with SS7. The SCCP provides routeing of signalling messages based on a global title (a globally valid address) which it translates to an MTP point code and SCCP sub-system (SCCP user) number. It also supports connectionless and connection-oriented services, and segmentation and re-assembly of higher-layer signalling messages. The combination of the MTP and SCCP is referred to as the network services part (NSP), which provides reliable transfer and global routeing and addressing of signalling messages through the network. Users of the SCCP are the ISDN user part (for some end-to-end signalling procedures) and the SS7 transaction capabilities.

To date, no protocols have been specified to support the transport, session and presentation layers of the OSI model in SS7. These layers are referred to as the intermediate services part (ISP), and are responsible for the end-to-end transfer and presentation of information to and from the application layer.

The transaction capabilities of SS7 are designed to interface directly to the OSI network layer, provided by SCCP in SS7. Currently, the only transaction capabilities are defined in the transaction capabilities application part of SS7 (TCAP). This is an application layer protocol that uses the SCCP connectionless service, and does not require an ISP. This makes the signalling system efficient in that TCAP messages do not have to be processed by ISP layers, but the flexibility that those layers provide is lost.

Other application layer protocols have been defined for use with SS7, and use the services provided by TCAP. These protocols, which include the intelligent network application protocol (INAP) and the mobile application part (MAP), are referred to as TCAP users.

3.4   SIGNALLING IN THE INTELLIGENT NETWORK

A feature of the intelligent network architecture is that it allows the platforms providing basic services (call processing) to be physically separate from the platforms providing supplementary services. Non-circuit-related signalling is required to convey information between the applications running on these platforms.

The approach taken to define the signalling for the new interfaces defined in the IN was to have a clear separation between the application-specific protocol and the underlying protocols that support it. This enables the application-specific protocol to be designed for high functionality and flexibility, and the underlying protocols designed for efficiency and resilience.

The network services part of SS7 (MTP and SCCP) fulfils the requirements for the underlying protocols; the OSI application layer (TCAP and TCAP users in the SS7 protocol reference model) fulfils those of the application-specific protocol. The TCAP user for IN signalling is the intelligent network application protocol (INAP). The INAP design methodology and protocol are described in the following section.

3.5   THE INTELLIGENT NETWORK APPLICATION PROTOCOL

3.5.1   The IN conceptual model

The IN conceptual model (see Chapter 1) defines four planes of decreasing abstraction — service, global functional, distributed functional and physical. These planes are used as the basis for the specification and design of the intelligent network application protocol to support services in an IN-structured network.

At the service plane, a target set of services to be supported are identified and described in an implementation-independent way, and decomposed into service features. The service features are realized in the global functional plane, the next level down, by global service logic. The global functional plane has an overall abstract view of the whole network and its capabilities. These capabilities are modelled as service-independent building blocks (SIBs), which are used by the global service logic to build services.

A number of SIBs are defined in IN capability set 1 (CS-1), covering areas such as charging, user interaction and call queuing. The basic call-processing capability of the network is abstracted as a single SIB, called the basic call process.

Functional entities within the network are identified at the distributed functional plane. This allows the functionality of the global service logic and SIBs to be distributed across the functional entities. In performing this distribution, functional relationships between the functional entities are defined. It is these functional relationships that determine the requirements of the signalling protocols used in the IN.

3.5.2 Information flows in the distributed functional plane

The distributed functional plane for IN CS-1 [6] defines a functional architecture comprising a number of functional entities and functional relationships. Part of this functional architecture is shown in Fig. 3.2. The functional entities are detailed further in Chapter 1. In the distributed functional plane, each of the SIBs are broken down into actions to be performed by the functional entities, and information that flows between them.

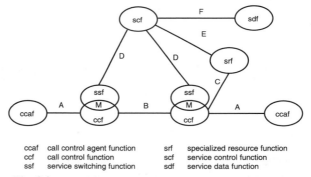

ccaf	call control agent function	srf	specialized resource function
ccf	call control function	scf	service control function
ssf	service switching function	sdf	service data function

Fig. 3.2   Call control and service-associated functional entities.

Each of the functional relationships between the functional entities (denoted in Fig. 3.2 by the reference points A to F) require control capabilities. These are:

- connection control (reference points A, B, C) — to establish, supervise and release a bearer connection;

- call control (reference points A, B) — to initiate and terminate the call, and provide the end-to-end control required for supplementary services which do not require IN service control;

- IN service control (reference points D, E, F) — to provide the control capabilities required for IN supplementary service control.

The distributed functional-plane architecture uses the call-control agent and call-control functional entities and the functional relationships (A, B) defined for these ISDN functional architectures [7]. These functional entities provide the non-IN call/connection control capabilities, referred to as basic call processing. This is modelled in the distributed functional plane as the basic call state model (see Chapter 1).

The only time that the IN service control interacts with the basic call state model is when armed detection points are reached in the call. When this occurs, information is passed to the service-switching function, which screens the information against the detection point criteria, e.g. whether the dialled digits match a particular digit sequence. If the criteria are met, then the service-switching function will initiate service control.

The non-IN call/connection control information flows (at the A and B reference points) have been defined as part of the ISDN, and form the basis of the ISDN signalling protocols. The information flows at the D, E and F reference points are defined as part of IN CS-1. The functional relationship between the call control and specialized resource functional entities (the C reference point) is considered to have only the connection-control information flows defined as for the A and B reference points. Due to the close coupling of the call-control and service-switching functional entities, the information flows at the M reference point are not standardized in IN CS-1.

The information flows at the D, E and F reference points are identified for each SIB. Figure 3.3 gives an example of some of the information flows and actions that make up the user interaction SIB. This SIB is used in cases where user interaction with a party is required in a service, e.g. to play an announcement or collect DTMF tones. In the example in Fig. 3.3, the SCF requires the CCF/SSF

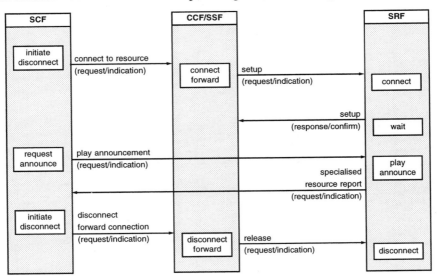

Fig. 3.3   Example functional entity actions and information flows in the user interaction SIB.

to connect the user to a specialized resource, and then instructs the specialized resource to play an announcement to the user. Once the announcement has been played — indicated by a report from the SRF — the bearer connection between the CCF/SSF and the SRF is released, unless further interaction is needed.

The complete set of information flows between any two functional entities, needed to support all of the declared SIBs, defines the functional relationship between those two entities. Some of the information flows are common to more than one SIB. For example, the connect to resource information flow is common to both the user interaction and queue SIBs.

3.5.3   The physical plane

The capabilities of the IN are realized in the physical plane of the IN conceptual model. Each of the functional entities are assigned to physical network elements, for example, as shown in Fig 3.4 (see also Chapter 1).

In some physical implementations, different functional entities may reside within the same physical entity (as in the SCP shown in Fig. 3.4). In these cases, the relationship between the functional entities is internal, and is not standardized. However, the INAP supports all of the functional relationships (D, E and F) as physical interfaces.

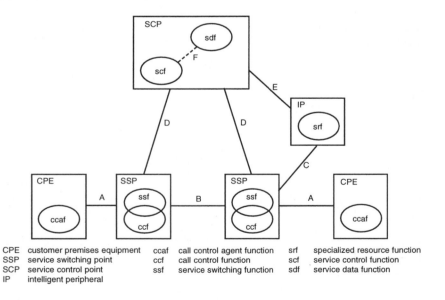

CPE	customer premises equipment	ccaf	call control agent function	srf	specialized resource function
SSP	service switching point	ccf	call control function	scf	service control function
SCP	service control point	ssf	service switching function	sdf	service data function
IP	intelligent peripheral				

Fig 3.4   Example physical plane showing allocation of functional entities to physical entities.

3.5.4   INAP protocol structure

The ITU-T recommendation Q.1400 [8] covers the development of signalling protocols based on the OSI application layer structure. This is a modular structure which enables the application layer to provide specific, expandable communications services to an application, while using generic underlying protocols. This structure has been adopted for the INAP, as it is ideally suited to the computer/computer communications required in the IN.

In the OSI environment, application processes (such as the SCF and SSF in the IN) contain communications application entities (AEs). These AEs are essentially the protocols and mechanisms that are responsible for all of the communications between peer applications. If an application process contains more than one application entity, it is the responsibility of the application process to control and co-ordinate them. For example, a service control function will contain different application entities to communicate with the service management function (typically a management-oriented protocol) and the service switching function (typically INAP).

An actual instance of an application process, i.e. running software, is referred to as an application process invocation (API). Each API will contain one or more application entity invocations (AEIs). This is illustrated in Fig. 3.5.

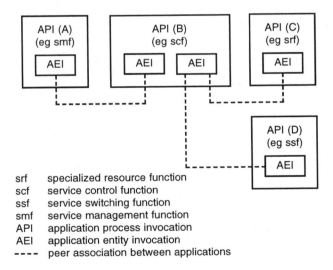

srf   specialized resource function
scf   service control function
ssf   service switching function
smf   service management function
API   application process invocation
AEI   application entity invocation
----   peer association between applications

Fig. 3.5   Relationship between APIs and AEIs.

The functional relationships between the applications are represented by application associations between an application entity and its peers. For each association, a single association object (SAO) exists within the AE which represents all of the communications capabilities required for that application association. These communications capabilities include the state models, operations, result reporting and error handling needed to support an association.

The communications capabilities are defined in related groups, referred to as application service elements (ASEs). The ASEs are governed within the SAO by a single association control function (SACF), which governs the necessary ordering and co-ordination of the ASEs. If more than one SAO exists within an application entity, the multiple application associations are governed by a multiple association control function (MACF). The structure of application entity invocations with one and multiple SAOs is shown in Fig. 3.6.

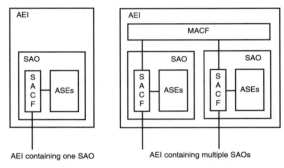

Fig. 3.6   Application entity invocation structures.

Not all of the ASEs within an SAO may be needed for a particular application association. An application context defines which ASEs and any related options that are required to support a specific application association, and is agreed by both peer application entities prior to an association being initiated.

The SS7 TCAP is an ASE which provides a framework for other ASEs and mechanisms for controlling an application association. It comprises two sub-layers — the transaction sub-layer and the component sub-layer. The TCAP transaction sub-layer is based on the OSI association control service element (ACSE [9]) and is responsible for establishing and releasing an application association between two application entities. Establishment includes defining the application context for the association and must therefore occur before any information is sent or received by other ASEs in the SAO.

The TCAP component sub-layer is based on the OSI remote operations service element (ROSE) [10]. This is a framework for defining remote operations in a distributed open systems environment. A remote operation is an operation which an application entity (invoker) requests another application entity (performer) to perform. The performer replies with the outcome of performing the operation, either a valid result or an error.

Remote operations supported by ROSE are classed by whether they are asynchronous or synchronous (i.e. whether the invoker may invoke further operations before the reply from the first is received or not), and whether success and/or failure or neither are reported. Linking together of remote operations may occur such that the performer of an initial operation may invoke further operations on the invoker of that operation, before the reply to the initial operation is returned.

These remote operations supported by ROSE must be invoked within an application association. They are supported in TCAP as simple structured or unstructured dialogues. Some enhancements to the ROSE protocol are included in TCAP, although only asynchronous remote operations are supported.

Figure 3.7 shows the structure of a service control function application process invocation in an SS7 environment.

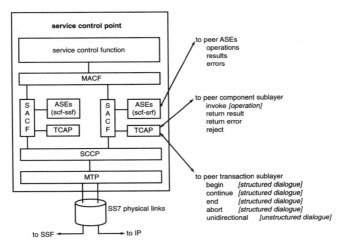

Fig. 3.7   SS7/INAP protocol model (at service control point).

3.5.5   INAP protocol definition

In the physical plane of the IN conceptual model, the information flows identified in the distributed functional plane are defined as remote operations consistent with the ROSE framework. The remote operations have optional arguments (corresponding to information elements), optional responses (results and errors), and optional linked operations which can be used as the response to a prior operation request.

The remote operations are grouped to form ASEs that define which entities invoke which operations. A number of ASEs relate to each physical interface, or SAO. The intelligent network application protocol is the collection of all of the ASEs and the necessary SACF and MACF rules, together with the definition of the actions taken at each entity to invoke and perform the required operations.

Since INAP is a ROSE user protocol it can be supported by any protocol stack that supports ROSE. This includes SS7 (TCAP) and DSS1. The INAP remote operations are defined in abstract syntax notation (ASN.1) [11], which makes them implementation-independent and portable to any processing environment.

Figure 3.8 shows an example of some of the operations between the SCF and SRF (which realize two of the information flows shown in Fig. 3.3). The 'play announcement' operation has an argument (PlayAnnouncementArg) and no defined results. This operation can be invoked by an SCF and is performed by an SRF. The argument will contain operation-specific information, such as identifying the announcement to play. In this example, two errors are possible (Cancelled and MissingParameter). These could occur if the operation is subsequently cancelled by the SCF, or if the SCF does not supply the correct arguments when invoking the operation. Since the 'play announcement' operation does not return any results, a linked operation (SpecializedResourceReport) is identified which can be used to return any information to the SCF about the execution of the 'play announcement' operation.

```
PlayAnnouncement:                           :=OPERATION
    ARGUMENT
        PlayAnnouncementArg
    ERRORS {
        Cancelled,
        MissingParameter
        }
    LINKED {
        SpecializedResourceReport
        }


PromptAndCollectUserInformation             ::=OPERATION
    ARGUMENT
        PromptAndCollectUserInformationArg
    RESULT
        ReceivedInformationArg
    ERRORS {
        Cancelled,
        ImproperCallerResponse
        }
SpecializedResourceReport                   ::=OPERATION
    ARGUMENT
        SpecializedResourceReportArg
```

Fig. 3.8   Example abstract syntax notation definitions of some SCF-SRF operations.

The 'prompt and collect user information' operation has an argument, result and two possible errors. This operation instructs the SRF to prompt the user to enter information, which is collected by the SRF, and returned to the SCF as the result of the operation. No linked operations are specified. The final operation, the 'specialized resource report', only has an argument specified (Specialized

ResourceReportArg), and is used to pass information (such as status reports) from the SRF to the SCF.

Figure 3.9 gives an example of an ASE to support specialized resource control. In this example, the ASE contains the three operations shown in Fig. 3.8. It defines that the supplier (in this case the SCF) invokes the 'play announcement' and 'prompt and collect user information' operations, and the consumer (the SRF) invokes the 'specialized resource report'.

```
Specialized-resource-control-ASE              ::=APPLICATION-SERVICE-ELEMENT
   --consumer is SSF/SRF
   CONSUMER INVOKES {
      specializedResourceReport
   }
   -- supplier is SCF
   SUPPLIER INVOKES {
      playAnnouncement,
      promptAndCollectUserInformation
   }
```

Fig. 3.9   Example INAP ASE (specialized resource control).

The INAP operations defined in abstract syntax are encoded into a concrete syntax (i.e. bit patterns) using the basic encoding rules [12]. The result is a set of encoded TCAP components which can be transferred to peer entities within a TCAP dialogue.

TCAP component sub-layer primitives allow the encoded INAP operations to be passed to TCAP for transmission, as well as providing an interface to control dialogues. The information relating to the operations, such as the name of the operation and arguments are carried in the component portion of a TCAP message. Information relating to the dialogue is transferred in the transaction portion. An optional dialogue portion can be used by the TCAP component sub-layer to specify and negotiate application contexts.

When supporting the INAP, the TCAP component sub-layer communicates with its peer using the INVOKE, RETURN RESULT (last), RETURN ERROR and REJECT component types. These component types are carried within the component portion of a TCAP message. The transaction sub-layer establishes and maintains a structured dialogue with its peer using the BEGIN, CONTINUE, END and ABORT message types, or supports an unstructured dialogue using the UNIDIRECTIONAL message type. These are carried in the transaction portion of a TCAP message. An example of message transfers between an SCP and an intelligent peripheral (IP), to request the IP to prompt a user to enter information (e.g. dial digits) and return them to the SCP, is shown in Fig. 3.10. In this particular example, both the SCF and SRF know that the dialogue will end after the result has been returned. Therefore, both ends only need to end the dialogue locally using a 'pre-arranged end'. This eliminates the need to send an 'END' message across the network.

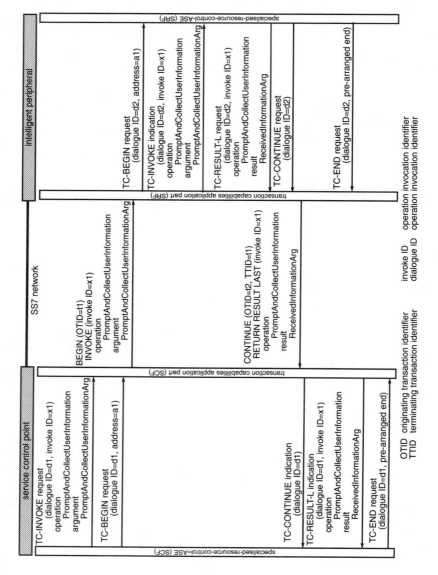

Fig. 3.10   Example message sequences between an SCP and IP.

ITU-T has defined the full set of ASEs required to support IN CS-1 [13]. In Europe, ETSI has decided to standardize a core set of these ASEs, referred to as core-INAP [14]. Application contexts are standardized for each of the functional interfaces and each of the various standards. The relationship between the application contexts, ASEs and operations defined in core-INAP is shown in Fig. 3.11.

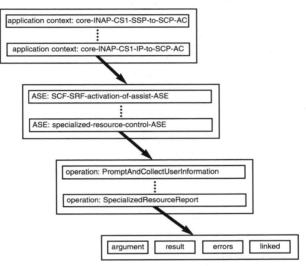

Fig. 3.11   Relationship between application contexts, ASEs and operations.

Addressing of the INAP application entities is performed using the SCCP addressing mechanisms (global title and sub-system number). Each application entity is identified by a global title, which is mapped to an SCCP sub-system number and MTP point code for routeing through the SS7 signalling network. Dialogues between application entities are identified in TCAP by a dialogue identifier. The ASEs required for the application association are identified by the application context.

3.5.6   INAP application entity procedures

INAP specifies the application entity procedures that relate to each of the INAP interfaces. These are specified as finite state models which define the procedures for using the SACF, MACF (if present), INAP ASEs (not including TCAP) and some of the functions of the application process invocation. The finite state models defined in the ETSI core-INAP contain a number of finite state machines:

- a functional entity access manager manages the sending, receiving, queuing and formatting of messages to and from the application entity finite state model;

- one management entity (or more) maintains dialogues with other application entities and performing operations which relate to management or supervisory functions (such as call gapping in the SSF); if more than one management entity is necessary, a management entity control finite state machine is specified — this is responsible for creating and terminating management entities and the call-associated finite state machines, and interfacing these to the functional entity access manager;

- one finite state machine (or more) performs the operations associated with single calls — these interface to the functional entity access manager via the management entity control finite state machine or the single management entity.

The structure of the SCF finite state model is shown in Fig. 3.12.

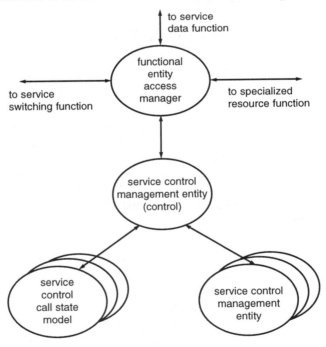

Fig 3.12   The structure of the SCF application entity finite state model (ETSI Core INAP).

3.6   SIGNALLING IN THE BT INTELLIGENT NETWORK

Support of the ETSI core INAP in BT's network has necessitated a major signalling network upgrade. This will enable SS7 non-circuit-related signalling, using SCCP and TCAP, as well as introducing the ISUP protocol for ISDN call control. The upgraded network is shown in Fig. 3.13.

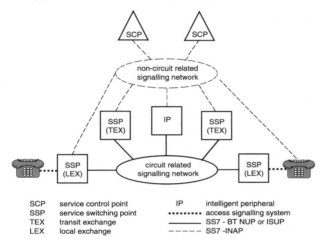

SCP	service control point	IP intelligent peripheral
SSP	service switching point	·········· access signalling system
TEX	transit exchange	———— SS7 - BT NUP or ISUP
LEX	local exchange	– – – – SS7 -INAP

Fig 3.13   The BT signalling network.

The use of SCCP and TCAP are primarily to support the ETSI core INAP protocol between SSPs, SCPs and IPs. To enable all SSPs (local and transit exchanges) and IPs to be connected efficiently to all SCPs, signalling point relays have been introduced into the SS7 non-circuit-related signalling network. These provide global title routeing of SCCP messages between the SSPs/IPs and the SCPs, and sub-system management. This not only reduces the number of connections required, but also increases the resilience of the network by providing routeing to an alternative (e.g. stand-by) SCP in the case of an SCP failure.

The BT network currently supports both analogue in-band access signalling (for example, pulsed and DTMF dialling) and ISDN access signalling (DSS1 and its own digital access signalling system No 2). Some of the IN-based services, such as number translation services, can be provided via all of these signalling systems. Others, particularly those which require user interaction, require the user to send DTMF tones. These services cannot be accessed from an ISDN terminal unless it is equipped with a DTMF tone generator. Services which currently use voice announcements, such as calling card services, obviously cannot be used with ISDN data calls.

3.7   SIGNALLING IN THE MOBILE INTELLIGENT NETWORK

Recommendations for signalling in public land mobile networks were published by the CCITT in 1988 [15], and were an attempt to define a signalling protocol which could support mobile facilities for voice and non-voice services. The mobile application part (MAP) of SS7 was developed based on the public land mobile network architecture and the procedures identified to support mobility (mainly roaming and handover).

In Europe, ETSI have standardized the global system for mobile communications (GSM) network, based upon the public land mobile network (see Chapter 2). This uses the MAP protocol for non-circuit-related signalling within GSM networks and interconnect between GSM networks. The MAP protocol has been further developed by ETSI, both in functionality and the interfaces supported. The GSM signalling network architecture is shown in Fig. 3.14.

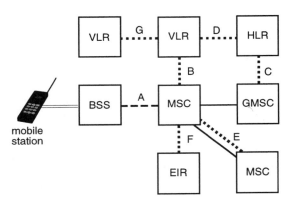

MSC	mobile switching centre
GMSC	gateway mobile switching centre
BSS	base station sub-system
VLR	visitor location register
HLR	home location register
EIR	equipment identification register

═════	RIL3 protocols
━ ━ ━	SS7 - RIL3 protocols and BSSMAP
────	SS7 - ISUP
••••••	SS7 - MAP

Fig 3.14   The GSM network signalling architecture.
(Note: the letters A-G indicate the functional interfaces. The H and I interfaces are not shown).

In the GSM network, signalling between the mobile station and mobile switching centre (MSC) uses the GSM radio interface layer 3 (RIL3) protocols.

These protocols are based on ISDN access signalling (DSS1) and support the call control, radio resource management and mobility management required by mobile terminals. Between the mobile station and base-station sub-system (the air interface), the RIL3 protocols are supported by a radio resource protocol. Between the base-station sub-system and the MSC, the RIL3 protocols are supported on SS7 by the direct transfer application part. Control of the base station sub-system from the MSC is performed by the base-station sub-system management part (BSSMAP) of SS7.

SS7 is used within the network for circuit-related and non-circuit-related signalling. Circuit-related signalling is required between MSCs and between MSCs and gateways to other networks. This can be provided by SS7 ISUP or operator-specific protocols, such as BT NUP.

Non-circuit-related signalling is required between MSCs, home location registers, visitor location registers and equipment identification registers and is provided by SS7 MAP. MAP interfaces are also specified between the mobile station and home location register (I interface) for invoking services, and between the MSC and a messaging network gateway (H interface), to support the GSM short-message service.

The MAP protocol is specified as a ROSE user protocol, and has operations grouped into ASEs to support each of the identified functional relationships. However, MAP does not conform to the application layer structure for signalling systems (ITU-T Q.1400), since this was not established at the time MAP was developed. The SS7 protocol model used for MAP is shown in Fig. 3.15.

No further development of the MAP protocol has occurred in the ITU since the initial 1988 recommendation. It is expected that any future network protocols to support mobility will be part of the INAP.

3.8   FUTURE DEVELOPMENTS

The use of the OSI application layer structure as a framework for the development of the INAP has resulted in an adaptable protocol that is well placed to take advantage of current and future trends in computing and telecommunications. As a ROSE user, the INAP can be used with any communications system that supports ROSE functionality, including current signalling systems such as SS7 and DSS1. Having a modular structure enables the INAP to be easily enhanced to support new capabilities and interfaces.

The next phases of IN standardization, capability sets 2 and 3, will place new demands on the INAP protocol. Targeted services — including universal personal telecommunications (UPT), global virtual network communications (GVNC) and broadband ISDN (B-ISDN) — will require new capabilities and new interfaces (e.g. SCP-to-SCP) to be supported within networks and across network boundaries. This may include the need to support INAP in the access

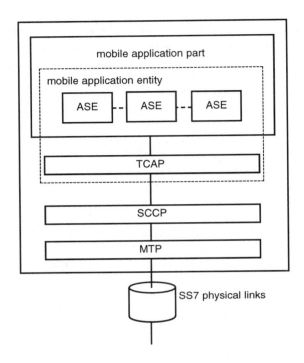

Fig. 3.15   The SS7 MAP protocol model.

network, for example, to support UPT registration from ISDN or mobile terminals.

It is expected that co-operation between the network and intelligent CPE in providing communications services will be a strong driver in the development of future intelligent networks. Extension of the INAP to the user via the ISDN access signalling channel (D-channel) would increase the user's ability to interact and control the network. Other aspects of ISDN, such as the use of service and terminal capability information to determine what type of basic service is being provided and how, also require further consideration.

The use of the intelligent network is applicable to B-ISDN, although alignment of IN and B-ISDN is not expected until IN CS-3. Broadband ISDN network signalling [16] has also been developed in accordance with the OSI application layer structure. However, while the B-ISDN functional architecture does not include service control, a functional separation between the call-control functionality (end-to-end signalling) and the bearer-control functionality (required to switch and control the bearer circuits) has been made.

This separation needs to be taken into account by the IN conceptual model, and will impact on the INAP, e.g. new information elements will be required in

INAP operations to support connection bandwidth and quality of service. New detection points in the basic call state model and service-switching functionality will need to be defined (e.g. to support number translation on terminal-to-terminal look-ahead messages, requests to modify bandwidth, or to control multi-party, multi-connection calls) which could result in new INAP operations and ASEs.

REFERENCES

1. ITU-T Recommendation Q.700, Introduction to CCITT Signalling System No. 7 (1993).

2. Fretten K G and Davies C G: 'CCITT signalling system No. 7: Overview', British Telecommunications Eng J, 7, Part 1 (April 1988).

3. Clarke P G: 'Internodal signalling', British Telecommunications Eng J, 13, Part 1 (April 1994).

4. CCITT Recommendation Q.931, Digital Subscriber Signalling System No 1 (DSS1) — general aspects (1988).

5. CCITT Recommendation X.200, Reference Model of Open Systems Interconnection for CCITT Applications (1988).

6. ITU-T Recommendation Q.1214, Distributed Functional Plane for Intelligent Network CS-1 (1993).

7. ITU-T Recommendation Q.71, ISDN Circuit Mode Switched Bearer Services (1993).

8. ITU-T Recommendation Q.1400, Architecture Framework for the Development of Signalling and OA&M Protocols Using OSI Concepts (1993).

9. CCITT Recommendation X.217, Information Technology — Open Systems Interconnection — Service Definition for the Association Control Service Element (1992).

10. CCITT Recommendation X.219, Remote Operations: Model, Notation and Service Definition (1988).

11. CCITT Recommendation X.208, Specification of Abstract Syntax Notation One (ASN.1) (1988).

12. CCITT Recommendation X.209, Specification of Basic Encoding Rules for Abstract Syntax Notation One (ASN.1) (1988).

13. ITU-T Recommendation Q.1218, Interface Recommendation for Intelligent Network CS-1 (1993).

14. ETSI European Telecommunication Standard ETS 300 374-1, Intelligent Network (IN); Intelligent Network Capability Set 1 (CS-1) Core Intelligent Network Application Protocol (INAP) (1995).

15. CCITT Recommendation Q.1051, Mobile Application Part (1988).

16. Law B: 'Signalling in the ATM network', BT Technol J, 12, No 3, pp 93-107 (July 1994).

4

MODELLING INTELLIGENT NETWORK SERVICES USING FORMAL METHODS

K C Woollard

4.1   INTRODUCTION

This chapter explores the problems associated with designing and specifying new services and the potential problems of interworking a number of services together. Furthermore it discusses the problem of interworking services across different network platforms, e.g. intelligent network (IN) based services interworking seamlessly with ISDN (integrated services digital network) services.

The chapter also explores the techniques available to enhance the understanding of a service during the critical stages of design and specification and in particular presents the advantages gained from rigorously modelling the service. The fundamentals of modelling are discussed with particular attention to the key issues of abstraction, focus, simplicity and clarity. The formal methods available for design and modelling are introduced with emphasis on the use of the Specification and Description Language (SDL) [1-3] and its associated tools. Using this technology it is possible to develop a good understanding of the operation of a single service and, when simulated in combination with others, it is possible to study the interaction between multiple services and the basic call control.

The ability to study the interaction between services is highly desirable, especially at the very early stages of service specification [4, 5]. In this way it is possible to highlight the potential interworking problems long before they are introduced into the network, where they degrade the quality of service and are costly to rectify.

Some examples of problems solved by employing these new techniques are detailed, together with how these are being introduced into the equipment procurement and testing activities.

4.2   SERVICE SPECIFICATIONS AND STANDARDS

The specification methods employed by BT, and others such as the ITU, have for many years relied on informal techniques such as English descriptions and message sequence diagrams. While these methods have been acceptable in the past, the problems that can arise from such a free format are beginning to be understood. In particular a number of telecommunications operators are reporting interworking problems within their networks.

These problems arise for a number of reasons. The main cause is that new services are developed in isolation from both existing services and others under development. For example, the ITU ISDN standards were developed in isolation from existing services such as BT's ISDN and centrex. This was partly due to the conflict of national versus international standards. The ITU has developed ISDN in isolation from mobile and IN services. To support these services a number of different, and incompatible, network solutions have been adopted by the telecommunications industry. As a result there are many, and varied, protocols both in the access and core network, and a number of similar but incompatible services. For example, it is possible to identify at least three different call-diversion offerings in BT's network relating to centrex, ISDN and network services. It would suit most network operators and customers if the services and protocols were compatible across all networks.

A further cause of interworking problems is associated with the methods used to define both existing and future systems. Because of the informality of the current methods it is not possible to understand either the precise behaviour of, or the implications associated with, new product definitions.

The need for new services is forcing the pace, with technology expected to deliver faster and faster service provision with a more and more complex set of services. This brings about an interesting paradox where, looking forward, it would be pleasing to see a multitude of new services, highly flexible service creation environments and numerous service providers, all seamlessly interworking. However, even with just a few services on the current networks, very little interworking has been achieved.

Because rapid changes are required in these areas, two specific scenarios need to be resolved. For current systems there is the task of identifying problems and solving these by modifying the service in operation. At the same time the requirements of building the vision of tomorrow without making the mistakes of yesterday must be faced. It is clear that there is a need to change working practices by, for example, developing all networks and services in a common context, and technology and methods are needed to help.

4.3   SEAMLESS SERVICES THROUGH RAPID MODELLING AND PROVING?

One of the major difficulties that exist in removing the current interworking problems is the limited understanding of the detailed operation of the present network and services.

To address this problem BT's service specifiers have started to adopt proven techniques from the product development arena. In particular by applying advanced specification methods, such as SDL, it has been shown that complex products can be developed at lower cost with improved quality. Furthermore, these products can be reused or partially reused in subsequent developments, and later modifications are easy to realize.

The products that can be developed using formal methods cover most of the software associated with the telecommunications environment, i.e. the switch control software, the protocols and the services. In general, these can be considered as separate products, since services in the future will be added with a degree of independence from the underlying network.

The use of SDL in the context of resolving existing network problems is relatively new in BT, but significant progress has been made in investigating services such as network services, ISDN and centrex. It has been shown that an SDL model of the service is able to independently detect known interworking problems such as the fundamental problems between call diversion and call waiting.

These two simple services typify the current problem. Call waiting and call diversion were both developed as network services and have been available on System X switches for some considerable time. It was not until there was sufficient geographical coverage of these switches that BT attempted to launch the services on a national basis. It was then, some time after development, that it was discovered that they did not interwork and could not both be offered to a customer. This is not only a problem for BT as examples abound in other telcos.

The main cause of the shortcoming in product development and testing was the lack of a sufficiently clear and complete specification. This specification has been retrospectively produced as an SDL model and, as a result, the cause of the problem has been identified.

Encouraged by these successes other services have been modelled with similar results. Known interworking problems have been reproduced and more importantly new interworking situations have been identified.

In particular, time has been spent on developing an IN service model as part of a feasibility trial of these methods.

Initially a number of issues associated with the specification of the services were identified. Mostly these were associated with the clarity and completeness of the specification.

Having developed the service definition, the formality of the language and the associated tools can be used to produce a simulation of the system being designed.

The simulation allows the design and behaviour of the services to be studied both in isolation and in combinations with other services. In particular, the service behaviour can be validated against the original requirements, allowing rapid and cheap redevelopment if it does not conform. Having validated against these requirements the service can be investigated against further sets of requirements such as feature/service interworking and possibly portfolio coherence issues such as the consistent use of tones and announcements. Furthermore, having developed a high-quality model, it is an ideal basis for specification, implementation and testing.

The use of SDL to develop service models is leading to the model-based development life cycle shown in Fig. 4.1. The current life cycle still exists in this model with the stages from informal requirements, through specification and development, to the product, and with the parallel path from requirements to test suite. However, Fig. 4.1 shows the addition of the SDL model and its importance at the beginning of the life cycle. In particular there is an iterative process taking place between requirements capture and model development that is essential to generating a clear set of requirements and a complete model. Once the model has been established it can be tested for conformance to the requirements and any inconsistencies can be corrected.

Once validated the model can be used in conjunction with the requirements to produce the specification of the product. In this case the use of SDL for the specification is shown and often there is a strong similarity between the specification and the model. Similarly, the model can be used to develop more detailed test cases to enhance those derived from the requirements. These tests can be produced in a semi-automatic way from the SDL model using the available SDL tools.

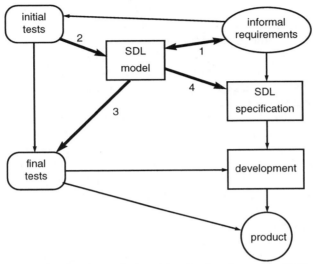

1. Iterative development of model and enhancing requirements.

2. Execution of initial test suite against model.

3. Development of final test suite based on the requirements and the model. Maybe automatically generated from the SDL.

4. Direct input from the model to the specification process.

Fig. 4.1   Model-based development life cycle.

4.4   FUNDAMENTALS OF A GOOD MODEL

The approach being advocated here is a combination between abstract specifications and pure modelling. Indeed both have their place. In the engineering world of complex systems the need for real models has long been accepted. For example, no one would dream of producing a new aeroplane without wind tunnel tests on a model. Similarly all new ships are subjected to extensive model testing in a wave tank.

The reason for extensive modelling in the engineering world is relatively clear. The systems being developed are complex and in some cases there are no sound scientific principles on which to move from specification to implementation with guaranteed success. The examples above are both cases of three-dimensional fluid flow for which there is no complete mathematical model. The opposite may be considered true of bridge building, where the laws of mechanics have proven reliable in the past — in this case the move from engineering drawing to the final product can be made in one stage with a predictable outcome.

In the software world of communications and computing there are very complicated systems which are growing in complexity all the time. Like the fluid-flow problem there is science to aid in the development of products. It seems sensible then to resort to the tried and tested principle of building and testing a model before implementation.

Like the model boats, those aspects of a system of interest need to be identified and captured in the models. For example, a toy yacht looks like the real thing, sails like the real thing, but is obviously not a real yacht. The toy has the essentials of a hull, keel, mast and sails. It may be constructed in a different material and is usually scaled down in size. It will also have some elements that perform a needed function that are different to the items found on a real boat. All of these issues add up to the concept of abstraction, i.e. representing the real item in a limited but meaningful way. Other issues that need consideration are completeness (does the boat have sails?) and simplicity (are all the features needed clearly visible and easy to use?). For any model to be useful these issues must be satisfactorily treated. One last point about the model boat is that the boat building industry relied for many generations on carefully constructed models for providing all the measurements for the real vessel. This represents the classic example of moving from requirements, to model, to implementation.

Developing specifications at the correct level of abstraction is essential to successful specifications. Too much information and the implementation choice is restrained, too little and the specification is incomplete, leading to divergent products.

In the IN, the services are separate from the underlying network and its protocols. As a result the abstract notion of a basic call is more important for the IN than previous systems. Having accepted the need to include a basic call within the service specification, an abstract model that is independent of hardware or protocols can be defined. The role of this basic call is to represent the behaviour of the network that would otherwise be missing from a purely service-level definition. The relationship between the basic call and its service models and the real network is shown in Fig. 4.2.

In this way the basic call is not intended for direct implementation, as is the case for the services, but rather it should have a traceable relationship to the more detailed network and protocol specifications associated with an implementation.

This mix of functional separation and abstraction within the specification provides the ability to focus clearly on the issues that need to be resolved in the model. In the definition of IN services the issues pertain to the customer-to-service relationship, service-to-network (basic call) and service-to-service. Each of these issues must be covered within the service specifications and each area can be refined into a more detailed set of issues.

- Customer-to-service relationship

  In the relationship between the customer and the service the concern is that the presentation of the service is consistent with other service offerings. This

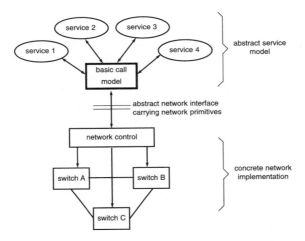

Fig 4.2   Relationship between service models and network elements.

is usually termed the 'look and feel' of a service and covers such things as the service invocation sequence (e.g. * # codes), the recorded announcements played and the way the service completes. If these elements of the specifications are not controlled, it is unlikely that the customer will experience coherence in ease of use.

- Service-to-network relationship

The area most covered in IN-based service specifications is the relationship between the services and the basic call. This is not very surprising as this represents the 'usual' operation of the network. Of course, this is an over-simplification, but, in general, it is essential to understand how the service is provided by the network.

- Service interworking

The third area identified as of interest is that of service interworking. This is currently a very active research area due primarily to the complex nature of the problem. In general, the services implemented are relatively few and simple, with limited interactions. However, telecommunications networks world-wide have started to experience interworking problems between services. As described earlier, even simple services have been found to be incompatible, requiring network upgrades. It is clear that if problems exist with current services, many more problems will be seen in a service-rich IN. The first line of defence is to improve the specification techniques by considering, modelling and hence managing the interworking of services.

For any specification a clear set of functional requirements is needed and the act of modelling helps to achieve these. For IN-based services, specifications of the service behaviour, interworking with the basic call, interworking with the customer and interworking with other services are needed.

By imposing the above considerations on model building, an attempt is being made to produce specifications that are focused on the important issues, are as simple as possible and hence as clear as possible, and, lastly, as complete as necessary to fully define the service.

4.5   METHODS OF MODELLING

Most of the above discussion has been generic and can be resolved by one of a number of specific methods.

It is possible for the specification and modelling activities to be separated, e.g. with specifications written in English and rapid prototypes developed in the software language 'C'. The problem with this approach is the high level of effort required to produce both a good English specification and a 'C' prototype. Furthermore, there is no semantic relationship[1] between the specification and the prototype. In fact this gives an abstract specification and an example implementation. The lack of a semantic link means that other implementations may still be demonstrated to conform to the specification, yet exhibit behaviour incompatible with the prototype.

A more ideal solution would be one with a very strong link between the specification and the prototype model. To provide such a solution the technology of formal languages has been developed. These formal languages strive to provide the best of both worlds, i.e. the expressive power of informal languages is needed to allow the descripton of things in an implementation-independent and abstract way. At the same time a strong semantic basis is required so that tool support can be developed and the designer can reason about the behaviour of the system based on the specification. The strong semantic basis allows the specification to maintain a strong link to both prototypes and implementations.

---

[1] The semantics of a language gives the precise meanings of the elements of the language. It is semantics that allows the extraction of the meaning from statements in the language. Formal languages such as SDL have a good semantic definition based on mathematics and it is possible to build tools to implement those semantics. A semantic relationship implies that there is a provable and reproducible relationship between two representations of the same thing.

The use of such an environment at last allows the addressing of the outstanding problems of incomplete and imprecise specifications.

Examples of formal languages include SDL, ESTELLE, LOTOS from the standards world, Z and VDM from the software world and a whole host of specialist languages from academia such as CCS and CSP.

Each of these notations has good and bad points and can be applied in different circumstances to different effect. The skill is in selecting the appropriate method for the requirements.

In the telecommunications industry there is a growing support for the use of SDL as a method for both specifying and implementing systems. There are a number of reasons why SDL is gaining wide acceptance:

- it has an easy-to-use graphical syntax;

- it is based on finite state machine theory, which is highly suited to distributed message-passing systems such as a telecommunications network;

- it has excellent tool support that allows full system analysis and validation of the specification;

- it is possible to automatically produce prototype models from the specification and, in a development environment, real code can be produced at the press of a button.

It is possible that in the future the use of formal specifications will be seen on the interface between BT and its suppliers. This could have advantages both at the procurement stage and during testing and integration.

There is a general problem of abstraction associated with all specification, modelling and implementation work. Certainly it is true that standards bodies such as ITU do not wish to produce specifications biased towards a particular implementation and hence manufacturer. At the same time specifications should not be so open that there is little chance of interworking. Currently there appears to be no panacea to this problem and each exercise relies on the judgement of technical experts.

Whatever the conclusion, the use of SDL should help the development process. In the case where the specifications are very detailed it is possible to consider code generation directly from the specification. This is an option being taken up by many of the product developers in the telecommunications industry with encouraging results.

If the specifications are very abstract, then the ability to simulate them allows the opportunity to clearly understand the behaviour required. It is this ability to model in an abstract plane that is central to the work on IN-based service modelling. The prime aims of this modelling work are to validate the service specifications and to detect any unwanted system behaviours, such as feature interworking problems.

4.6   MODELLING INTELLIGENT NETWORK SERVICES

A number of activities have been carried out within BT to assess the usefulness of SDL for specification, modelling, and implementation. The findings for implementation are clear and are detailed in Cookson and Woodsford [6] and are echoed by the findings of others in the industry.

In all the specification work undertaken it has been found that the act of producing a formal specification has posed a lot of questions about the original set of requirements, and led, in general, to moving from an informal set of specifications in English and Message Sequence Charts, to a formal specification in SDL. The formal specification is invariably more complete and of a higher quality. This means there are less opportunities for implementors to produce the wrong product, the development time should reduce and there should be fewer interworking problems.

SDL has been applied to the specification of a number of BT's proposed services with a number of interesting results. Detailed service design documents were accepted as input to the process. From these documents an independent set of SDL specifications was produced. The SDL specification demonstrated a high degree of correlation to the detailed designs indicating that the informal documents were of good quality.

Once the formal model was complete it was primarily used to look for unwanted feature interactions within the services. This was a manually driven simulation of various service combinations. In this way, it was possible to investigate the user's perspective of the individual services, the interworking of the services as a single package and the interworking between new services and existing network services.

The investigations highlighted 14 potential issues that may affect service interworking (see Table 4.1). All of these issues related directly to deficiencies in the specifications. Some would most probably be picked up during product development and should be reflected back into subsequent issues of the specifications. Others arise from areas outside the scope of the existing specifications. In particular, a number of problems identified were in the interworking between the new service and existing network services not considered in the service design.

These independent services are seen as the most problematic as they are likely to be developed by different product development teams and launched on to a platform with limited thought about their interworking issues. Furthermore, as the number of services and the service complexity increase, it will become more difficult to cover all the interworking issues in an informal way. Consequently, the results in this area are of particular interest as the modelling work offers a method of studying interworking issues at the specification level.

Table 4.1   Service modelling sample results.

Modelled Network Feature	Modelled Service			
	Chargecard	Personal Numbering	Remote Access to Call Divert	Remote Access to Call Minder
Calling number delivery	Fail	Pass	N/A	N/A
Call return	Fail	Fail	N/A	N/A
Call waiting	Fail	Fail	N/A	N/A
Three-way calling	Pass	Fail	N/A	N/A
Call divert immediate	Pass	Fail	Pass	N/A
Call divert on 'busy'	Pass	Fail	N/A	N/A
Call divert on 'no reply'	Pass	Fail	N/A	N/A
Call back when free	Pass	Fail	N/A	N/A
Number translation	Pass	Pass	N/A	N/A
Callminder	Fail	Fail	N/A	Pass
Chargecard	N/A	Pass	N/A	N/A
Personal numbering	Pass	N/A	N/A	N/A
Service access	Pass	Pass	Pass	Pass

Pass: indicates no interworking problems found.
Fail: indicates potential problems found.
N/A: means no tests are applicable.

It is beginning to become accepted that service modelling will be a necessary requirement for all services before launch. It is essential that service modelling is carried out as early as possible in the development process.

The findings from the studies and the application of the method on a wider range of products are currently being assessed. One of the areas that is already affected by the findings is product testing [7, 8].

Currently BT tests new products on an end-to-end service basis, making the assumption that satisfactory conformance testing was performed by the manufacturers. During the testing of new products it is essential to uncover as many problems as possible and the issue of service interworking is becoming paramount. Consequently, the results from service modelling are being used as the basis for testing the real product. There are two good reasons for such an approach:

- firstly, the results from the model are focused on the areas of the specification that are weakest and hence most prone to arbitrary decisions during manufacture;

- secondly, the interworking examples are too complex to be defined and results to be predicted in the usual (manual) way.

In these complex areas the services can be investigated as models and the results captured as message sequence charts. These charts can then be used to guide the testing and act as the expected test results.

4.7   INTERNATIONAL PERSPECTIVE

There is currently a lot of activity within other companies and international fora concerned with the application of languages, such as SDL, to specification and modelling. In particular there is a clear focus on the new technology areas, such as IN, and accepted problem areas, such as feature and service interworking.

There are activities within Eurescom (project P230 and proposal PP 5302/5) using SDL to address feature interworking in the intelligent network. Similarly the ACTS programme (the European initiative on Advanced Communications Technology and Services) is proposing work in the area of modelling and feature interworking.

Some suppliers can be seen to be making a strong move towards the use of SDL as a natural part of the product development life cycle. For the product developers the long promised 'holy grail' of automatic code generation is becoming a reality with informally quoted improvements in both coding efficiency and the number of errors produced in the code. Accepted figures for lines of delivered code vary between 100 and 400 per man month using conventional methods. From projects within BT, figures nearer to 5000 delivered lines per man month can be seen, when SDL is used as part of the development. This is an order of magnitude improvement.

Furthermore, the use of SDL as the design language has improved the link between designs and implementations, with the documentation being maintained as products are developed and enhanced. Manufacturers are also beginning to identify project management improvements from the use of SDL as the product can be split up, developed and then integrated in a controlled way.

There is also evidence that other telcos, in the UK and elsewhere, are pressing to use SDL on the interface with their suppliers.

Most of the advantages apparent to our suppliers and competitors are consistent with the findings from BT's work. It is becoming more evident that formal specification, modelling and implementation can significantly contribute to the quality and timeliness of product developments. Furthermore, the power associated with formal method tools can be used to help contain some of the complex problems such as feature interworking.

4.8   CONCLUSIONS

The use of formal models has been shown to offer advantages in a number of areas. For the first time there is the opportunity to document the capabilities of the current networks allowing new products, or services, to be assessed against existing products. In this way BT has direct control of the product portfolio coherence across the services on offer. Also, this provides a much enhanced understanding of requirements from manufacturers, reducing the 'confusion resolution time' currently associated with equipment contracts.

The opportunity exists both to assess the delivered product rigorously against BT's requirements and to maintain a tight control on the quality of network services.

It is true that the methods described in this chapter will add extra effort to the front end of the development life cycle, but benefit can be gained from all the advantages with the added bonus that overall product development times should be reduced.

REFERENCES

1.   ITU-T Recommendation Z.100 and annexes A, B, C, E, F1, F2 and F3 (1992).

2.   Saracco R, Smith J R W and Reed R: 'Telecommunications systems engineering using SDL', North-Holland (1989).

3.   Belina F, Hogrefe D and Sarma A: 'SDL: with applications from protocol specification', Prentice Hall (1991).

4.   Bryce K, Crowther M and King J: 'Feature interworking detection using SDL models', Globecom 94, San Fransisco (1994).

5.   Crowther M: 'Modelling and validation of telephone network signalling', SDL'93, Using Objects, Darmstadt (1993).

6.   Cookson M and Woodsford S: 'Design methodology using SDL', BT Technol J, $\underline{11}$, No 4, pp 16-24 (October 1993).

7.   Kelly B, Webster N and Bruce G L: 'Future testing methods using SDL', BT Technol J, $\underline{11}$, No 4, pp 25-34 (October 1993).

8.   Kelly B, Webster N, Boullier L, Phallipou M and Rouger A: 'Evaluation of some test generation tools on a real protocol example', 7th International Workshop on Protocol Test Systems, Japan (November 1994).

5

SPEECH SYSTEMS

K R Rose and P M Hughes

5.1   INTRODUCTION

The perceived quality of the services offered by a telecommunications company, and thereby the perceived quality of that company is, in part, derived from experiences which customers have when using or encountering network and voice services. This is particularly true with the complex services delivered by an intelligent network (IN). In particular, such perceptions tend to be based upon the ease of use and naturalness of these services. As such, it is important that voice automation and voice services are used in appropriate circumstances and are designed and structured so as to improve perceptions rather than lower them. (See the Appendix for definitions of terms used in this chapter.)

Managing customers' expectations is a subject that all companies wrestle with on a day-by-day basis and a comprehensive treatment of this subject can be found in Atyeo and Green [1]. However, a brief review of the human factors aspects of telephone network services is presented here to illustrate some of the problems facing designers of speech services.

By way of an example, consider making a telephone call to a mail order company. In this case, the customer would expect the call to be answered quickly and would wish to be dealt with in a polite and efficient manner. Even if the mail order company in question cannot provide the requisite service, they would wish to leave the customer with a good impression, so facilitating repeat business. Most companies strive for this goal when dealing with customers on the telephone, be it via human or automated operators. However, in practice, it is clear that the customer's expectation of an interaction with a machine is different to that of interacting with a person. Imposing a badly designed automated service or insisting that callers interact with automated services in inappropriate circumstances can ultimately be damaging to the business. Conversely, appropriate use of well-designed and implemented services can significantly enhance customer and caller perceptions.

In many cases, given the choice of interacting with a person or a machine, callers would not choose to interact with an automated service. An illustration of this is the response which many people have when confronted with an answering machine. In some cases, the caller will simply hang up and not leave a message. As the caller may be disappointed at not having made direct contact with the person they have called, this behaviour is accepted, but is difficult to explain rationally. In addition, the owner of the answering machine has installed the answering machine not only for their own convenience but also for that of the caller.

Although this is a simple example, it demonstrates that customers' expectation and attitude on meeting an automated speech system differs considerably to that of a person-to-person interaction. As such, a key problem facing developers of all automated voice services is how to manage the interaction so that callers are positively encouraged to continue the dialogue with the speech system, despite a possibly negative attitude to such interactions. This problem is all-pervading and at the current time is relatively independent of the specific technology or architectural approach used to deliver the network or speech service.

5.2   SPEECH SYSTEMS — A FUNCTIONAL OVERVIEW

A speech system in a telephone network has three principal functions:

- to provide general voice guidance, e.g. informing a customer when they have dialled an unrecognizable telephone number;

- to provide customers with automated control of network services, e.g. invoking a call-diversion facility;

- to provide voice services, such as call completion, voice messaging, information provision (e.g. Directory Enquiries), etc.

A speech system can be broken down into a number of key components which are shown in Fig. 5.1.

The nature and function of each of these components is discussed below.

5.2.1   Telephony interface

The telephony interface can be divided into two functional areas — the speech interface and the signalling interface.

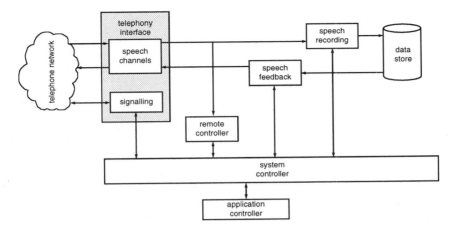

Fig. 5.1   Speech system functional block diagram.

In systems with relatively few lines (up to ten, say), the telephony interface will usually be an analogue connection. The signalling available from an analogue interface is limited to a small number of call progress tones, e.g. equipment busy tone or number unobtainable, which the speech system has to automatically detect in order to manage the call.

When the number of lines becomes large it is far more economical to use a digital telephony interface. This brings the added benefit of improved speech quality. Signalling over digital connections comes in two forms — channel-associated and common-channel. Although channel-associated signalling is much quicker and more reliable than analogue signalling, the call-management information is essentially the same as for the analogue interface. With this form of signalling, the analogue supervisory tones are allocated digital codes which are then embedded within the speech channels and can be detected and interpreted by the speech system.

However, common channel signalling, where the signalling and speech information are transmitted separately, contains extensive call-related data in addition to basic call progress information. This additional information can be used by a speech system for improving the quality or ease of an interaction with customers and callers. One example of such information is calling line identity (CLI). This can be used by the speech system to identify the customer, prior to any interaction with the speech service taking place, and allow customer specific information to be accessed, thus improving the perceived quality of the interaction. For example, the speech system might inform the customer which network facilities were currently enabled on their telephone line.

5.2.2   The remote controller

The purpose of the remote controller is to detect the control signals sent from the customer's telephone. The customer has two methods available for remotely controlling a service using the telephone — the keypad or dial, and the microphone, i.e. using speech.

The telephone dial or keypad were designed purely for call set-up and hence signalling from the telephone to the local exchange. With the advent of Touch-Tone™ signalling it was quickly realized that these tones could be transmitted across an established telephone connection (after call set-up) and reliably recognized at the receiving end. In the UK, however, there are some telephones that use loop-disconnect signalling, especially in the residential market. As a result, and in order to provide a ubiquitous service, it is necessary to consider the use of loop-disconnect transients[1] as a means of remotely controlling network services; however, the speed of transmission and the accuracy of recognition for loop disconnect transients is lower than that for TouchTone signalling.

In operation, the remote controller recognizes the signals sent by the customer's telephone and transmits them to the application controller (see section 5.2.7). The number of different signals that can be originated by a telephone dial/keypad is limited to twelve — the digits '0 to 9', '*' and '#'[2] — although loop-disconnect telephones usually transmit the digits only.

An alternative approach to providing callers with the ability to control network services from their telephone is to use voice recognition. In this case, the customer speaks appropriate commands into the telephone microphone. Using speech commands in this way has two principal advantages over the keypad/dial:

- there is, in theory, no limit to the size of vocabulary that is available;

- control of the service can be made to appear more natural and, therefore, easier.

---

[1] A telephone produces loop-disconnect signals simply by making and breaking the direct current on the local telephone line. During call set-up these direct current pulses are terminated at the local exchange and are not transmitted any further. If these pulses are transmitted once a call is set up, a series of transients is induced on the line beyond the local exchange and these can be detected at the receiving end of the telephone connection.

[2] There are 16 TouchTone tones; however, only 12 of these are used by most telephones.

To facilitate this type of control, the speech system must be able to recognize the speech commands spoken by the customer. Until recently the quality of speech recognition technology was not perceived as being acceptable for deployment in network services. This perception was reinforced to a degree by an early approach to the use of speech recognition, which was to simply replace TouchTone key presses with the spoken form (i.e. rather than dialling/pressing the digit '1', the customer was invited to say the word 'one'). This particular approach proved to be less than satisfactory for a number of reasons. Not only are the spoken digits notoriously difficult to recognize automatically — they appear to a recognizer to be very similar acoustically — but the control of the service was slower than using the keypad/dial. As a result customers became frustrated with the early voice-driven services.

The utilization of speech recognition as a form of remote control has, however, progressed substantially in recent years through advances both in the performance of the underlying recognition algorithms and the price/ performance of the hardware needed to run these computationally intensive processes. As the technology has advanced, the application of speech recognizers has tended to drift away from being simply digit detectors to that of assisting users to invoke facilities in a more natural fashion. To this end, speech recognition now tends to be used primarily to identify service command words.

An example of where the speech recognizer can be used in this way is in the customer invocation of network services. Currently, a customer invokes a network service, such as call diversion, by entering a unique code made up of TouchTone digits. However, customers may forget the code for the network service they wish to use. Using a speech recognizer, the customer could simply speak the name of the service required, e.g. 'divert', and the speech system would transmit the appropriate service code to the network. The keypad is used only for entering telephone numbers, e.g. destination telephone numbers for diverted calls. If the customer does not have a telephone with TouchTone signalling then the speech recognizer can be used for recognizing digits or, alternatively (but with less reliable results), loop-disconnect transient recognition may be used.

In order to optimize the accuracy of the speech recognizer the number of words to be recognized at any one time needs to be kept as small as possible and the words themselves need to be easily distinguishable. As such, allowable key words that are used within a service need to be carefully chosen to ensure optimum performance. In the same way, the prompts or guidance messages given to the customer need to be worded carefully to guide the customer, politely and unobtrusively, to use the appropriate words.

5.2.3   Speech feedback

Speech feedback to the customer can be in the form of supervisory tones, speech (via the telephone loudspeaker), or visual indication (by utilizing the display that can be found on some telephones). Supervisory tones are still widely used for feedback to the customer, e.g. dial tone, busy tone. However, speech feedback is increasingly being deployed to improve the customer interface, e.g. the BT telephone network now plays a message to a customer when it does not recognize a dialled number instead of playing 'number unobtainable' tone. Visual feedback is increasingly being used as telephone instruments evolve to include display capability; this particular form of interaction is relatively new and is not considered further in this chapter.

The quality of the voice used in speech feedback is an important factor in creating a favourable impression with customers. Quality is both subjective and nebulous, however, a few basic attributes can be identified. The intonation of the voice should create an impression of helpfulness and politeness, and yet be assertive when required. For example, it may be necessary at some point in the dialogue to instruct the customer to undertake a simple action. The pace of the voice should be such that the customer does not feel rushed. The type of voice selected can also set the perceived 'style' or 'personality' of the company providing the service. For example, a Scottish company may purposely choose a Scottish accented voice for its automated speech services.

5.2.4   Speech recording

Some applications may necessitate the storage of customer information. However, speech requires large amounts of storage — for example, one minute of speech recorded at the standard digital pulse code modulation (PCM) telephony transmission rate of 64 kbit/s requires 480 kbytes of storage. At this rate the speech storage soon becomes a limiting factor with respect to cost and the time taken to retrieve the data. As a result, most speech systems employ some form of data compression to reduce the amount of speech storage required. The most common compression rate currently being used in commercial speech systems is 32 kbit/s[1], i.e. one half of the standard digital telephony transmission rate. Using speech compression also helps reduce the internal data transmission requirements within a speech system, ultimately helping to reduce costs.

---

[1]   The most common data compression technique used is Adaptive Differential Pulse Code Modulation (ADPCM).

5.2.5 Data store

Most applications require the storage of customer information. This information is usually in two forms:

- service information, e.g. what services the customer currently has available — this is sometimes referred to as the customer profile;

- application information, e.g. in a call-answering application a caller may leave a speech message for the customer.

This information is ideally suited to the ordered storage methods offered by standard database structures, e.g. relational databases.

5.2.6 The system controller

The system controller has two principal functions:

- to provide a simple means of managing the functions of the speech system;

- to undertake housekeeping to ensure that the speech system is functioning optimally.

5.2.7 The application controller

The role of the application controller is to co-ordinate the elements of the speech system in order to execute a defined transaction, or dialogue, with the customer/caller. All the functions described above (i.e. speech feedback, speech recognition, TouchTone recognition, etc) can potentially be used for interacting with a customer. The only differences between the dialogues for different applications are:

- the order in which the functions are implemented;

- the feedback messages given;

- how the recognized TouchTone tones/loop-disconnect pulses are interpreted;

- the words to be recognized;

- customer or caller information that is to be recorded and stored.

The applications controller makes requests to the systems controller to perform some task, e.g. record the customer's voice.

Having reviewed the functional aspects of a speech system the remainder of this chapter is dedicated to describing BT's latest network-based system, the speech applications platform (SAP), and how it has been designed to meet BT's requirements for supporting future telephone network-based voice services.

5.3   THE SAP DESIGN PHILOSOPHY

The speech system that has been deployed into the BT telephone network for supporting new voice services, such as CallMinder™, is the BT speech applications platform (SAP). The SAP has been designed to function in both telephone network and private system environments. Although the SAP equipment used in these environments is similar, each imposes its own set of operational requirements.

At the outset of the SAP development an underlying design philosophy was adopted to try and ensure that a number of critical success criteria were met. This underlying approach and a description of the system structure is given in the following sections.

5.3.1   Scaling

The principal objective for a telephone network designer is to provide services to the customer at optimal cost. As part of the telephone network, the speech system should also be designed to meet this objective. One significant method of cost optimization is to share the available resources between as many different applications as possible.

To achieve this objective the SAP was designed to be as modular as possible, not only at the macro level from a functional unit perspective, e.g. application processor, but also at a micro level within each functional unit. This is described more fully below.

5.3.2   Modularity

Any large system such as the SAP can be architecturally divided into smaller functional units. The SAP is split into six processing units:

- telephony;

- signal processor;

- customer management unit (file-store);

- system management;

- system back-up and data collector;

- local area network (LAN).

The function of each of these units is described later in the chapter. Each unit functions autonomously, hence the SAP has a distributed processing architecture.

Communication between all units is via the LAN. Additional units can be added to the system to meet service requirements and the system reconfigured to optimize cost. If, for example, a new service were to be launched where a SAP installation had spare customer management unit (file-store) capacity, then it would only be necessary to add the appropriate signal processor units required to handle the additional traffic.

The functional units are modular in design and can be configured to suit the service requirements. Each unit has a finite amount of processing capability, and yet the processing requirements for each service differ. In order to match the system to the processing needs of the application, the configuration of each unit can be adjusted to optimize the usage of processing resources and hence optimize the cost of providing the service. This technique of configuring resources to suit the service is known as 'dynamic resource allocation' and is described further in the section on cost optimization (see section 5.3.3).

Finally, being autonomous processing units means that development on each can take place independently. For example, the telephony unit can be enhanced to exploit a new signalling protocol without affecting the other units in the system.

5.3.3   Cost optimization

The ultimate performance goal of any service-providing system is to supply the requisite amount of resource to meet a pre-specified level (quality) of service. Providing too much resource incurs a cost penalty and providing too little means a reduction in the level of service. For example, telephone exchanges are designed to meet a specified grade of service which will dictate the maximum number of calls that will be handled in the busiest hour of the busiest day.

The majority of speech platforms are designed such that there is a fixed amount of speech processing resource assigned to each telephone channel. If the service that runs on the platform does not require all the assigned resource then there is a cost penalty. At the other end of the spectrum, the maximum amount of resource available to a service is limited, which can limit the services that can be supported.

To overcome this limitation the resources on the SAP are treated as a single pool and the requisite amount of resource is allocated to a service as required. During the execution of a service, the amount of resource required will depend on the function being executed. For example, the playing of a speech feedback message requires only a small amount of resource compared with executing a speech recognition function. Using this approach, only the required level of resource need be made available to execute any particular function. The net effect is that the resources are employed to the maximum efficiency and thus the cost of the service is optimized. This technique, dynamic resource allocation, is one of the features that distinguishes the SAP from other speech systems available today.

5.3.4   Industry and open standards

Industry standard components and open standards are employed on the SAP wherever possible providing the following advantages:

- the market for industry-standard components is very active and very competitive, resulting in a wide range of products being available to system developers — the diversity of the market means that a single source of supply is rarely a problem and that prices are keen;

- companies in the market are continually developing new and improved products with better price/performance;

- the companies in the market will develop new products to meet the latest standards.

The net effect of all the above points is that system designers have at their disposal a number of cost-effective, high-powered computational building blocks that can be used to build an optimum system from both a cost and performance perspective. Upgrade capabilities are provided regularly through the market-place, giving a straightforward future-proofing route.

5.4   THE SAP TECHNICAL SOLUTION

As noted previously, the SAP consists of six processing units. Figure 5.2 shows a functional block diagram of a SAP network element for supporting new voice services such as the BT CallMinder call answering service in the BT telephone network. The nomenclature in italics in Fig. 5.2 corresponds to the functional units described in section 5.2. This system has been connected into 21 sites in the BT telephone network primarily at digital main switching units (DMSUs). The system is based on the industry standard VME Bus which is used for transmitting control information between the various elements within each processing unit. The function of each processing unit within the SAP is described below.

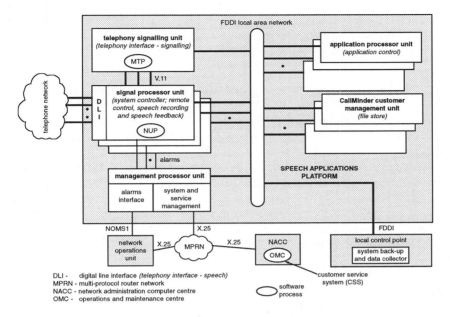

Fig. 5.2   The BT speech applications platform network element.

5.4.1   Telephony signalling unit (telephony interface — signalling)

All telephony communication, both speech and signalling, is provided by standard 30-channel PCM systems, terminated on the digital line interface on the signal processor unit (see section 5.4.3). Each channel provides a 64-kbit/s bearer which in the case of speech is companded using A-Law encoding (note that the telephony signalling unit handles the signalling functions of the telephony interface).

The signalling channels are separated from the speech channels on the digital line interface. The signalling information is then transmitted to the telephony signalling unit (via a V.11 interface). The signalling employed is ITU-T common channel signalling system No 7 (C7). The telephony signalling unit handles levels 2 and 3[1] — message transfer part (MTP) — of the signalling protocol and extracts the level 4[1] call processing information — network user part (NUP). The NUP contains the call processing functionality required to carry out the network service, e.g. status of line, CLI. The NUP is transmitted to all the signal processor units via the FDDI LAN.

---

[1] These C7 signalling levels do not correspond to ITU-T open systems interconnection (OSI) layers.

The telephony signalling unit contains three signalling links from separate parent telephone exchanges — this arrangement is sometimes referred to as triple parenting. This is to enable alternative network routeing in case of traffic congestion or exchange failure.

5.4.2   Application processor (application control)

The function of the application processor is to run the applications software. As mentioned earlier, each application is a dialogue transaction with the customer/caller.

An application process manages the telephone call via the telephony signalling unit and makes requests to the signal processor unit for signal processing functions via an object-oriented application programmer interface (API). The API is a library of high-level functions that allow an application to be developed while at the same time shielding the application developer from the underlying complexity of the SAP system. For example, the application processor could issue the request 'play stored message X to telephone channel Y' without any knowledge of the type of telephony interface or the compression rate at which the message was recorded.

The application processor can be located either in each of the signal processing units, or as a separate central resource that communicates with each signal processing unit via the LAN, as is the case in the service node architecture (see section 5.5).

For clarity, Fig. 5.2 shows the latter configuration; however, the SAP systems integrated into BT's telephone network have the signal processor units and application processors collocated on the same shelves.

5.4.3   Signal processing unit

The signal processing unit performs the roles of system controller, remote controller, speech feedback, and speech recording. This unit is shown in Fig. 5.3.

As the system controller, this unit is effectively the heart of the SAP and is divided into five main modules:

- **digital line interface** (telephony interface) — this unit terminates all digital telephony links and separates the signalling information from the speech data; most of the data channels carry digitized speech signals and are transmitted to the signal processors via the SpeechBus;

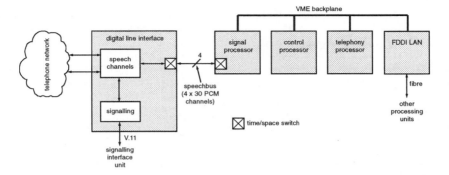

Fig. 5.3   Speech processing unit functional block diagram.

- **SpeechBus** — it is possible to transmit speech data over the VME control bus; however, as mentioned earlier, digital speech signals have a high data rate and the available bandwidth on the control bus would quickly be exhausted; this would limit the amount of control information that can be transmitted with the result that fewer voice services can be supported per system — as such, and like most large voice systems commercially available today, the SAP transmits control information and speech data on separate buses; as the name suggests the SpeechBus is used for transmitting the speech data, leaving the VME bus for control information only — the SpeechBus has four 30-channel PCM highways that are used for transferring speech data between the digital line interface and the signal processors, connection to the SpeechBus being via time/space switches located on the digital line interface and signal processor units; this provides the capability for the speech data to be switched internally within the SAP and creates the infrastructure required for dynamically allocating the speech resources;

- **telephony processor** —this processes the C7 signalling NUP call-processing information from the telephony signalling unit and converts this information into a form that can be manipulated via the application programmer interface (API);

- **control processor** — this has two primary tasks: to manage the signal processing resources when requested by an application (via the API), and housekeeping; on receipt of an API request the control processor schedules the speech processing resources to undertake the task, switching the speech data from the digital line interface to the appropriate signal processor units via the SpeechBus — as housekeeper, the control processor continually monitors the state of all modules to ensure that they are functioning correctly;

- **signal processors** — these execute the various signal processing functions (e.g. speech recognition, speech recording, speech feedback) requested by the control processors.

5.4.4   Customer management units (data store)

These units are the mass-storage systems which contain the customer profiles and customer speech data for different applications. In Fig. 5.2 the customer management unit holds the customer data for the call minder call-completion service. The industry standard interfaces network file system (NFS) and small computer systems interface (SCSI) have been adopted for these units so that the appropriate technology can be utilised to meet the service requirements. In the case of the BT telephone network implementation, fixed magnetic disk technology has been used with each individual disk having a 'shadow' for resilience. The customer management units are shared between all the other units and the data is managed by using a relational database management system.

5.4.5   Management processor

This unit is the interface between the SAP system management and the BT network management system. It has three principal functions:

- to raise alarms to the network operations unit (NOU) via the network operations management system (NOMS) interface;

- to undertake system management commands requested by the network engineer in the NOU, e.g. the engineer may interrogate the status of a particular unit if an alarm is raised — these commands are transmitted from the engineer's system console in the NOU via the BT X.25 multi-protocol router network (MPRN)[1] with the statistical information, such as the number of calls handled, also provided upon request from the network engineer;

- to update customer information, e.g. add a new customer — this information would be transmitted from the customer service system (CSS) and collected by the operations and maintenance centre (OMC) software in a network administration computer centre (NACC) and transmitted on to the SAP via the MPRN.

All the SAP processor units can send and receive operational information, such as statistics, alarms, etc, to and from the management processor which then converts this information into the format required by BT's network management systems.

---

[1] Until recently, this X.25 network was known as the administration data packet network (ADPN).

5.4.6   System back-up and data collector

As the name suggests this unit has two roles. As the system back-up unit it holds an up-to-date software image of the whole SAP system together with the applications software. Periodic back-ups are undertaken so that, in the case of failure, the system can be restored to the known state at the time of the last back-up. This back-up is stored on a magnetic tape cartridge.

As a data collector the unit can be programmed to monitor and record transactions undertaken on the SAP. For example, if the performance of the speech recognizer for a particular application needs to be assessed, then all the invocations of this application will be recorded by the data collector. This information is then stored on some form of bulk removable storage media, such as optical disks, and removed for analysis. The system back-up and data collector is usually located in the same place as the telephone exchange back-up tapes, e.g. local control point (LCP).

5.4.7   Local area network

All the above units communicate via an FDDI optical fibre LAN, although for smaller installations, an ethernet LAN is used.

5.4.8   Software overview

From a software perspective the SAP is a collection of independent software components each providing an integral part of the speech system. From a technical perspective these units have considerably different constraints placed upon them and the approach to the software design reflects this difference. In line with the design philosophy, either open and/or industry standards were adopted as appropriate.

Although there are many software components, the key decision made for the software design was the choice of operating systems. Two operating systems were adopted:

- UNIX™ — for supporting the applications;

- VxWorks™ — a commercially available real-time operating system that is very similar to UNIX in design and used for supporting the SAP control software.

A full description of the SAP software is not appropriate to this chapter; however, to give a flavour of the software operation a single API command process is described.

Figure 5.4 shows a diagrammatic representation of the software processes involved in the execution of a typical API command.

The application processor is running the UNIX operating system, whereas the remaining processors are running the VxWorks real-time operating system.

For example, assume the application program issues the API command, 'Recognize_TouchTone_KeyPress'. The API passes this command to the application control process (that runs on the control processor) which, in turn, requests the resource for this function from the resource allocation process.

The resource allocation process reviews the current state of the signal processes (running on the signal processor unit) in the available pool and decides which of these should execute the TouchTone recognition — this is the heart of the dynamic resource allocation process adopted on the SAP. Several factors are taken into consideration when making the decision; for example, is a signal process resource available that has just executed a TouchTone recognition so that it can be reused and therefore avoid having to re-configure another resource?

Once the resource has been identified the resource allocation process informs the speech control process (that runs on the signal processor) on which process to run, on which resource to run the process, on which SpeechBus channel (not shown in Fig. 5.4) to expect the incoming tone, and to which application process the recognized tone should be returned.

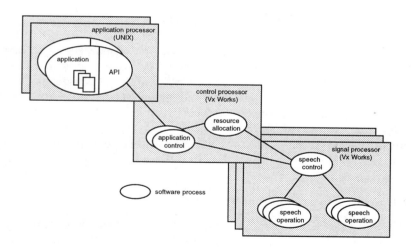

Fig 5.4   Typical SAP software operation.

The TouchTone recognition process is then performed (speech operation) (it should be noted that, in practice, many speech operations are being undertaken simultaneously). Once the process has been executed, the result (the recognized TouchTone digit) is passed to the speech control process and then on to the application control process. The speech control process also informs the resource allocation process that the signal processing resource used for the TouchTone recognition is now available.

Finally, the application control process returns the result of the TouchTone recognition to the application program via the API. The system operation is very complex owing to the large number of concurrent processes; however, it is this concurrent operation that provides optimal use of the available resources.

5.5   THE SAP AS A SERVICE NODE

The BT service node architecture is covered in more depth in Chapter 6. Effectively, the only SAP component used in the service node architecture is the signal processor unit. All signalling and customer data handling is performed by other parts of the service node. The level of application control undertaken by the SAP is also determined by the service node.

5.6   THE SAP AS AN INTELLIGENT PERIPHERAL

It is planned that by 1996 the current SAP nodes in the BT telephone network will be modified to function as intelligent peripherals (IPs) as well as a 'stand-alone' speech systems. Figure 5.5 shows the modified SAP architecture that includes the IP functionality. The principal modifications lie in the telephony interface and how applications are managed.

- telephony interface — the telephony interface is modified to run the C7 signalling connection control part (SCCP) as well as the message transfer part (MTP), the combinations of these software functions being known as the network services part (NSP) (this upgrade gives the telephony interface compatibility with layer 3 of the open system interconnection (OSI) model);

- IP service unit — the IP does not require a CMU and instead an IP service unit is provided; this unit supports the C7 transaction capability application part (TCAP) and specialized resource function (SRF) to enable communication with a remote applications processor, known as a service control point (SCP); service data, e.g. speech announcements, is stored as part of this unit.

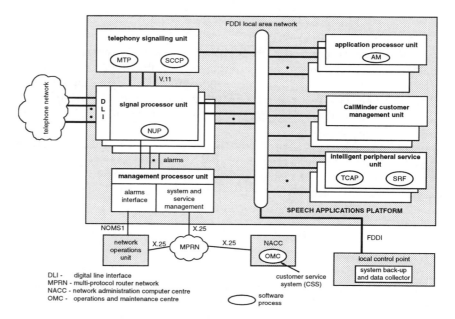

Fig 5.5   The BT SAP as an intelligent peripheral.

For an IP application the applications manager (AM) (which runs on the application processor unit) will pass control of the application to the SCP which is usually located remotely from the SAP installation. The SAP communicates with the SCP using the intelligent network application part (INAP) SRF.

For a non-IP application the AM will interact with the local CMU to service the call and in doing so instruct a signal processing unit to execute the appropriate application.

5.7   THE SAP SPEECH PROCESSING FUNCTIONALITY

The following processing functionality is currently available on the SAPs integrated into the BT telephone network:

- TouchTone recognition — recognition of TouchTone digits transmitted from the customer/caller;

- loop/disconnect transient recognition — recognition of loop/disconnect transients transmitted from the customer/caller;

- TouchTone generation — use of TouchTone digits to control remotely other items of telephony equipment, e.g. pagers;

- speaker-independent speech recognition — the recognition of up to ~ 30 words spoken by any person — would be used for the command words of a network service; the smallest vocabulary would consist of the words 'yes' and 'no' which would be used for very simple control of voice services such as call minder;

- speaker-dependent speech recognition — the recognition of up to ~ 30 words spoken by a particular person — would be used when a customer has chosen specific words for a particular application, e.g. repertory dialling;

- speech recording and playback at 64 kbit/s — the facility for speech data to be recorded and played back at the standard telephony PCM rate, used when speech storage requirements are minimal;

- speech recording and playback at 8 kbit/s — the speech is recorded at one eighth of the standard telephony PCM rate (64 kbit/s) which offers a major saving in data storage over other commercially available speech systems; the quality of the reproduced speech is at least as good (subjectively) as the speech feedback from telephone answering machines;

- silence detection — this is used to detect when customers/callers have stopped speaking when recording speech;

- call progress tone detection/generation — used for signalling on analogue telephony circuits;

- modems — used in applications that require data transmission, e.g. remote meter reading.

It is planned to upgrade the SAP to include a number of new and advanced functions. These new functions will considerably expand the range of services which the SAP can support and further enhance its usefulness to BT.

5.8   SAP TOOLS

Several tools have been developed to assist in using the SAP and are outside the scope of this chapter. However, the Visage application generation tool [2] should be mentioned in that it has significant benefits for the developer of network services. Visage is a PC-based tool with a graphical user interface that assists with design of voice services. The tool imposes a set of criteria upon the designer to ensure that the service developed meets those criteria defined in a style guide [3]. By using the tool it can be ensured that each service has the same 'look and feel'. Once the service has been designed it can be directly loaded on to the SAP application processor via the FDDI LAN. Therefore, the time taken to develop and test new services is dramatically reduced.

5.9   CONCLUSIONS

Speech systems are an integral part of the intelligent network and can provide the sophisticated functionality necessary to control the advanced services offered by the intelligent network. However, the key to the successful take-up of network services will be the design of the customer interface. As the services become more and more sophisticated, the challenge faced by human factors design engineers grows. It takes time for customers of such services to learn to control the new features and to interact with services in a new or different way. As a result it may be that this learning time, and not technology, will determine the rate of take-up of these new services.

APPENDIX

Definitions

For the purposes of this chapter the following definitions have been adopted:

- customer — person who wishes to implement a network service (e.g. call diversion) or owns a network voice service (e.g. voice mail);

- caller — person who is met by a network or voice service when making a telephone call;

- telephone — the method by which a customer remotely controls access to a network service via the telephone network; control is through the use of speech or call 'set-up' signals, i.e. loop-disconnect transients or dual tone multi-frequency (DTMF) tones, often referred to as TouchTone;

- network service — telephone management service, such as call diversion, message waiting, etc;

- voice service — automated customer voice services, such as call answering, voice messaging, etc.

REFERENCES

1.   Atyeo M and Green R: 'User friendly weapons for the competitive fight', British Telecommunications Eng J, 13, Part 3, pp 201-205 (October 1994).

2.   Hanes R G et al: 'Service creation tools for creating speech interactive services', EuroSpeech, Berlin (1993).

3.   'Now you're talking ... voice services', BT Document (1994).

6

THE SERVICE NODE — AN ADVANCED INTELLIGENT NETWORK SERVICES ELEMENT

S Kabay and C J Sage

6.1   INTRODUCTION

With the accelerated trend towards more open and competitive global telecommunications networks, many traditional network operators are rapidly losing their share of the lucrative mass telephony market to new entrants. As a result of technological advances and regulatory controls, users now have a choice of communications products and services available to them.

In order to retain large users and win back customers from new entrants, network operators are being forced to provide greater flexibility in configuring their networks to make best use of available resources and provide the rapid deployment of new and innovative services.

The challenge for the major network operators is to overcome the inherent architectural limitations of a switch-based legacy network. At present, national service deployment in a switch-based network requires all switching manufacturers to develop the same features in a transparent way. Due to the logistical and commercial issues associated with such an upgrade, the process is costly and time consuming.

As an alternative, many network operators have invested in overlay networks to provide flexible new services to leading edge customers. While improving responsiveness to market forces in the short term, overlay networks result in fragmented access and management of the overall network.

Due to these difficulties, network operators have opted for the long-term aim of developing a core intelligent network (IN) capability. The concept of the IN provides major advances in the provision of services to the user while allowing the service provider increased ease and flexibility in deployment. The IN standards and architectural issues are discussed in Chapter 1.

Having identified the IN as the target infrastructure that will provide the competitive edge in an advanced telecommunications market, the network operator must address the non-trivial task of migrating services from an embedded switch-based PSTN to a core IN architecture. The key changes that must be made include:

- increasing the signalling functionality of the PSTN to handle the additional traffic associated with IN-based services;

- adopting an open systems interface to support service-independent signalling protocols between the various IN elements.

The latter must conform to international standards and is of critical importance if IN-based services are to operate transparently across equipment provided by multiple vendors.

Successful migration from the existing switch-based services to a distributed IN architecture will only be possible following a detailed analysis of the various network constraints. These include minimizing service development costs, ensuring the least possible impact on existing PSTN resources, and allowing the smooth introduction of services with gradual traffic growth. In addition, possible evolution in services and the likely emergence of unforeseen demand for some services must be considered.

Working in collaboration with researchers at BT Laboratories, who have been involved in shaping the future architecture of the BT core IN and the evolving international standards within CCITT and ETSI, the network intelligence engineering centre embarked on a development programme to produce a self-contained IN element known as a service node. The main objective of the service node (SN) development was to deliver, within extremely short timescales, a network-worthy platform to trial certain IN-based services within the PSTN. Recognizing that the core IN will take some years to realize, investment was made in the SN to gain early insight into the technical and architectural issues associated with the implementation of IN.

Based on an open system architecture, an SN is ideally suited for assessing the feasibility of many IN concepts and services by rapid integration of emerging best-of-breed technologies. By performing trials of certain IN-based services on a selected customer population within a live PSTN environment, this study provided BT with the ability to prototype and field trial IN-based services and concepts in advance of the core IN development.

This chapter discusses some of the technical and architectural issues identified during the development of an SN into a network-worthy advanced IN-based services element. The deployment of the SN to trial certain IN-based services is also discussed. The chapter concludes by exploring the possible application and reuse of SN technology within BT's core IN.

6.2   KEY REQUIREMENTS

Some of the key requirements adopted during the design of an SN were based on the following prerequisite characteristics of an IN element:

- extensive use of information processing, efficient use of network resources;

- modular, reusable and scalable;

- conforms to standardised, service-independent protocols and communications;

- multivendor capability;

- multinetwork capability;

- rapid service delivery;

- service deployment.

This SN design philosophy has resulted in a generic framework for supporting advanced services. By adopting existing, established computer technology to provide the underlying control and processing capability, an SN architecture readily supports an open interface which allows for the ability to procure and integrate commercially available, third-party network resources. This approach is consistent with the IN objective for rapid service creation and minimal re-engineering of network elements to support new services.

A major requirement is that an SN should support a generic service-independent framework, whereby new resources (e.g. interactive voice response, voice recognition, facsimile store and forward, screen phone, audio-conferencing, e-mail) can be easily integrated to provide additional service capabilities.

The latter implies an inherent flexibility to support an easily scalable and modular design. This functionality would enable BT to meet the specific needs of its corporate customers by providing specialized services, possibly with a short life expectancy, in support of a customer's business strategy.

In the likely emergence of unforeseen demand for some trial services, an SN should support a strategy which allows the timely roll-out of the service post-trial, while extending the customer base of the service. Alternative scenarios can be depicted where an SN is deployed to support extended marketing trials or to provide BT with the ability to support a selective marketing strategy during a transition to the core IN environment.

Many other requirements that impacted upon the design of the SN architecture included performance engineering issues, continuous availability, high throughput, real-time call-handling, dynamic reconfiguration, platform scalability, service independence, and the ability to support multiple network interfaces.

6.3 FUNCTIONAL PROTOTYPING AND SYSTEM SPECIFICATION

This section summarizes the results of the rapid prototyping phase of the SN development. This exercise was used to validate, verify and refine SN requirements. The main objectives included:

- verification of the SN functional requirements;

- a detailed definition of the SN software and hardware requirements;

- development and evaluation of a functional SN prototype to determine quantitative and qualitative benchmark and design criteria.

Based on previous studies [1], the system architecture adopted a distributed multiprocessor architecture based around the UNIX™ operating system. For improved performance, each of the major sub-systems had to be designed to operate efficiently in a multitasking and, where available, multiprocessing environment.

A key requirement of the control processor, which forms the core processing capability of an SN, was the ability to prioritize and schedule application/network requests in real time. This capability was essential if a predetermined quality and responsiveness of service was to be offered to customers.

Significant performance benefits could be attained by writing customized memory management functions to override the generic functionality offered by the operating system. By optimizing memory performance, the control processor could minimise the level of context switching between SN processes, thereby increasing overall efficiency and throughput.

The decision to adopt UNIX as the underlying operating system is based on the results from previous evaluation studies [1, 2]. The key benefits of UNIX, within a multiservice telephony environment, include widespread multivendor support, ease of portability of applications and the flexibility of having supplier-independent target platforms.

However, a number of limitations were identified in the UNIX kernel during the performance benchmarking of the SN prototype. The main problem was the single-threaded design of the UNIX scheduler, which adversely affected initial estimates of system performance. Further tests on a range of UNIX-based products showed this to be a common limitation to all of the systems bench-marked.

Despite these problems, the UNIX kernel is more than adequate for the level of processing required to achieve current market predictions for services. How-ever, these problems have been notified to the relevant suppliers and this bottle-neck should be eliminated in future releases of UNIX.

6.4   ARCHITECTURE OVERVIEW

At the core of an SN architecture is a network database much like a service control point (SCP), but it also contains a transport interface and peripherals, similar to an intelligent peripheral (IP), which can be used to facilitate user interactions with the network. Figure 6.1 illustrates the SN architecture.

A brief summary of each of the hardware components is given below:

- control processor — comprises a fault-tolerant commercially available computer system running the UNIX operating system;

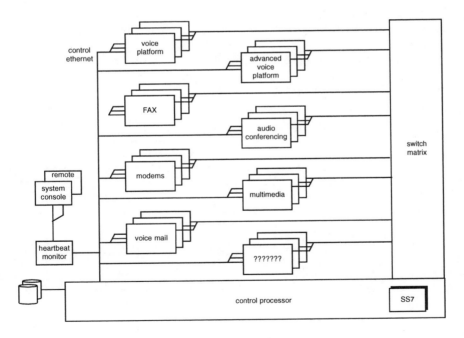

Fig. 6.1   The SN architecture.

- local area network (LAN) (multiple ethernet) — a typical system configuration requires a minimum of two ethernets for security, and can currently support a maximum of eight;

- resource/switch interface — European primary rate multiplex hierarchy standard E1 (USA standard T1 also in development);

- system console — a maximum of four consoles, which can be located locally or remotely, can currently be supported;

- CCITT signalling No. 7 (SS7) — the current SS7 implementation is closely coupled with the control processor hardware;

- switch — the switch currently supports 256 30-channel PCM systems, multi-frequency detection being supported on every channel (a number of exchange tones are also provided by the switch as a generic capability);

- resources — resources are provided using an $N+1$ sparing policy.

The SN switch is a core component that serves as a concentrator to provide access to a family of resources. The resources can provide such voice-service functions as dual-tone multifrequency (DTMF) detection, speech synthesis, speech recording and playback, voice recognition and a broad range of other facilities.

Figure 6.2 shows the major software sub-systems of the SN.

The key components of the control processor will be described in later sections.

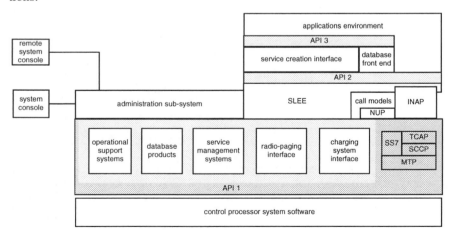

Fig. 6.2   Control processor overview.

6.5   CONTINUOUS AVAILABILITY

A key customer requirement is that an SN should provide continuous availability. This characteristic has a major impact on the overall cost of the platform. Consequently, very careful consideration was given to the level of resilience and fault tolerance required of the various processing elements that make up an SN. Certain criteria were devised to compare the mission-critical components against economic factors and performance.

The mission-critical components within an SN were identified to be:

- control processor — controls the service logic and platform resources;

- network interface — handles the lower level SS7 signalling;

- switch matrix — handles the routeing of signalling and speech into the platform;

- inter-process communications — enables the reliable transport of messages between the control processor, switch and SN resources.

The required system behaviour of these components leads to a system architecture that must embody hardware reliability and flexibility. The SN provides continuous availability by incorporating a combination of (high-cost) hardware fault tolerance and (low-cost) $N+1$ redundancy techniques.

The SN control processor (and integrated network interface) is based upon a commercially available, fault-tolerant hardware architecture and operating system. Despite the high capital cost, this approach greatly simplified the design effort required to implement the sub-systems that were resident on the control processor. The designers of the SS7, service logic execution environment (SLEE), SN accounting and charging (SNAC), and node manager sub-systems, could rely on the underlying hardware environment to sustain system availbiity.

The SN switch matrix adopts a combination of hardware fault tolerance and $N+1$ sparing to provide high availability.

A critical element of an SN architecture is the internal LAN which carries the inter-process communications between the various sub-systems. This mechanism was implemented using a dual-ethernet configuration to provide the necessary resilience. In addition, a new software protocol was developed above the UDP/IP stack to provide assured LAN reliability, performance and responsiveness.

The SN architecture limits the effect of a single point of failure such that it does not propagate throughout the system to adversely degrade performance. This characteristic is further enhanced by designing a secure resource allocation and control mechanism within the SLEE. Consequently, a given resource can only be accessed through a well-defined interface, which guarantees that any such access will not corrupt nor cause failure in any other SN component.

6.6   HUMAN FACTORS IMPLICATIONS

The service node architecture and functionality is defined by a set of service-independent functional building blocks (FBB) distributed across the SN and its resources. The scope of each FBB was designed to be fully configurable while limiting the capability of each block. Typical FBBs include: detect incoming call, receive digits, store digits, analyse digits, route call, check call status, play announcement, perform simple speech recognition and free speech circuits. To accommodate a wide variety of services of different complexity, the FBB concept provides highly flexible and reusable software components with specific service-independent capabilities. Each FBB has set input parameters that specify the required functionality, e.g. these could either include the text to be played to the user during a voice prompt, or specify the structure of any information required as input from the calling customer.

The rationale of the FBB was to reuse existing building blocks on the SN resources, thereby reducing software development and maintenance costs. This approach is fully consistent with the IN objective of rapid service deployment.

The main service that was identified for benchmarking the SN is known as personal numbering (PN). This service provides the customer with the ability to register a series of physical telephone numbers to a single, logical personal number. Each customer has a personal profile maintained within the SN database which can be updated on a daily basis to register the desired time-of-day routing. The customer is also given a default number to which calls can be routed should they need to deviate from their planned itinerary. An integrated voice-mail system and a number of ancillary support services were also provided to assist with customer mobility.

The PN service provides a sophisticated call-completion facility, whereby the calling party can hunt up to three customer-configurable telephone numbers (on busy or no-answer) before calls are terminated to voice-messaging.

The interface provided to the PN customer when accessing the service to either configure their profile, or to access the personal number service feature set, took the following form:

- initially the customer is given a greeting and requested to enter their account number and PIN for authorization (this is designed to provide security and fraud limitation to BT and its customers);

- if the customer is successful, the SN will play the following announcement: "Good <time of day>. You have <number> new voice message(s)";

- the customer would then hear the main menu for the service.

Using the FBBs, this entire prompt could be requested using a single FBB request to the speech resource using parameters to specify each speech segment within the prompt.

6.7   THE NETWORK INTERFACE

The SN control processor contains a set of integral fault-tolerant network interface cards which are used to perform portions of the SS7 signalling protocol — in particular, the message transfer part (MTP), signalling connection control part (SCCP), and transaction capabilities application part (TCAP) layers.

The SS7 capability of the control processor was of major importance in the selected implementation strategy. In addition, the SS7 had to support open and stable interfaces. Using the defined interfaces, customized user parts could be developed that could be transparently integrated into the SN environment.

A key development that had to be undertaken was BT's national user part (NUP), which handles PSTN call set-up and call processing. The NUP provides direct circuit-related signalling and control. Ongoing developments include the implementation of ETSI capability set 1 (CS-1) IN application part (INAP) and the mobile application part (MAP) protocols. CS-1 INAP will provide the SN with the capability to perform connectionless IN-related functionality. The MAP will provide the ability to interconnect with the cellular networks, so that the SN can provide transparent multinetwork access.

The SN has the capability to support multiple application user parts to allow easy interconnect to a host of other licensed operator (OLO) networks.

Basic link management and configuration functions have been integrated into the SN system console via the node manager sub-system. The approach taken for the integration of the SS7 sub-system is typical of the approach adopted for the integration of third-party products throughout the development programme.

6.8   THE RESOURCE INTERFACE

The SN provides a consistent and secure method of integrating new resources into the node. By adopting a layered SLEE, the service logic was isolated from the SN resources and lower-level basic call processing. By abstracting the service logic from the underlying platform architecture, new service features can be easily added without having to upgrade any of the core SN software sub-systems.

The SLEE can be regarded as a real-time call-processing system, autonomously handling basic telephony and providing an open interface to additional features for supplementary services. The application is a purely sequential program that has a completely synchronous, transaction-oriented interface with the SLEE. In this way, self-contained applications can be readily developed, the functionality of which is via system calls to the SLEE.

The SLEE keeps track of and detects events and conditions that require an application instance. An application is usually triggered either by an event or a condition at a predetermined point in a call. Once a trigger is detected, the SLEE will schedule the application to take control of a call.

The remainder of this section will present an example of service requiring a customer interaction with a speech resource. The example will be limited to the message flows over the interface shown in Fig. 6.3.

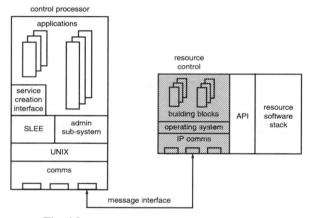

Fig. 6.3   Control processor resource interface.

- The application makes a SLEE API_Reserve_Resource request to play an announcement and receive DTMF digits. The application is allowed to specify the name it will use to reference the resource; thus, the application programmer can adopt their preferred resource naming convention within a specific service.

- The SLEE will then issue an API_Connect to the switch requesting that the incoming call be connected to the specified resource.

- The selected resource must also be instructed to perform the required function. In this example, the resource must play an announcement and receive digits. The SLEE constructs the API command, using the appropriate parameters, and despatches the request to the resource.

- Once the caller is connected to the resource, the SLEE will issue a Start_Dialogue to the resource which will then play the selected announcement.

The SLEE philosophy provides application developers with a platform-independent method of constructing services, minimizing the development effort required to construct a service.

A further benefit of the SLEE is that it minimizes any degradation in performance due to inadvertent faults in SN resources. For example, if a particular resource is causing a processing bottleneck, the SLEE resource allocation algorithm can be used to smooth over such difficulties to deliver an acceptable level of responsiveness to the customer.

6.9   FLEXIBLE ACCOUNTING AND CHARGING

A key objective of IN is the capability to allow the service provider to maintain control of routeing decisions for services, based upon time of day, day of week, authorization codes, etc. The routeing decision criteria will be under the control of the service provider. Similarly, charging decisions can be based on locations, destinations, time of day, authorization codes, etc. The service provider will require similar flexibility to define the desired charging criteria for every aspect of a service.

The SN accounting and charging (SNAC) system is the centralized service node system that produces platform-independent usage records (PIUR) containing customer service usage data. The application developer is provided with a flexible mechanism for generating:

- charge records for customer billing and invoicing;

- statistics data for marketing analysis and platform auditing.

The SNAC data generation function (distributed across the application, SLEE, and call model) produces raw customer service usage data when specified generation criteria have been fulfilled. Once a complete service-independent usage record (SIUR) has been built, the SLEE will forward the SIUR to the SNAC for storage, formatting, and outputting to the relevant user.

The SNAC provides the platform with the capability to charge for the following generic charge events:

- incoming call events — based on the duration of a call to a specific platform feature or resource;

- outgoing call events — based on the duration of an external call to a specific destination or platform;

- elapsed time events — based on the duration of a call while using a specific platform feature or resource;

- one-off events — based on the selection of a specific platform feature or resource, for example, activating a radiopager.

Many other parameters, including absolute timings, geographic routeing, type of resource used, etc, will also be collected by the platform and made available to the billing system, such that the tariffing and costing algorithm used to calculate the actual cost can take into consideration all data associated with a call.

In order to support IN-based services, the SNAC can cope with a wide range of charging and accounting data depending on the type of resources accessible to a service. It is not sufficient to simply generate new call-record structures to support new services.

Specifically, the SNAC accounting capability can generate:

- customer and service profiles, which can be used by a marketing service for assessing the efficiency and effectiveness of the service or platform;

- detailed customer service usage that occurs during a single instance of a service, resulting in a complete account of usage;

- aggregate customer service usage across multiple instances of service that have been provided within a specified aggregation interval;

- usage data formatted with full or partial call-records and output to the destination billing system.

The SNAC has the level of flexibility to support new services and any associated call-record formats on an evolutionary basis.

6.10   INTERCONNECT

The cost-effective realization and timely delivery of services are essential requirements for network-deployed IN platforms. Hence, the IN platform must be adaptable to ensure interconnection and communication with a broad range of existing network elements. The SN adopts a flexible interfacing technique between key sub-systems such as the SLEE, SS7 NUP, SNAC, and specialized on-line customer databases. This allows the SN to rapidly 'graft' interfaces to new platform resources in a very short time interval.

To achieve this, a generic conversational resource interface was designed into the SLEE. This enables the easy introduction of new interfaces with minimal disruption. This interface is extremely flexible and currently supports interfaces to a radiopaging centre, service management console, and cashless call validation/

billing database. More recently, development has commenced on the SS7 INAP and MAP protocols taking full advantage of this interface.

6.11   DISCUSSION AND CONCLUSIONS

This chapter has reviewed the issues associated with the design, development, and deployment of a service node. The main objective of this project was to deliver an advanced IN services element for trialling certain IN-based services and concepts within a live network.

Many issues have been identified during the development of the SN and the subsequent trial of selected services. Having demonstrated that an SN architecture can fulfil the main objectives of IN, the service trials have provided valuable insight into the issues associated with deploying IN.

The fundamental IN architecture depicts a fully distributed core network comprising SSPs, SCPs and IPs. The interconnect and messaging between these entities is via SS7 signalling links. It is envisaged that, for a typical IN-based service, such as a cashless services call, there will be in the order of ten or more complex INAP operations per transaction. Assuming an average SCP capacity of 500 transactions per second, during periods of heavy loading, this architecture will result in a significant amount of SCP-SSP-IP messaging. The likely observable outcome is an added signalling delay and customer perceived degradation in service. Overall performance and call throughput is further compromised by the additional processing overhead due to formatting, timing of responses, and verification of the received information.

The SN architecture overcomes this potential bottleneck by tightly coupling the SCP and IP functional entities into a single physical network element. Although the SCP-IP protocol remains conformant to the CCITT standards, the transport layer is based on high-speed UNIX inter-process communications. Using the accounting capability of the SN, future studies can be used to investigate the performance engineering issues associated with this messaging interface.

A major area of concern for many network operators is the potential loss of revenue due to fraudulent users. Many operators now consider fraud detection as a major system requirement. Preliminary studies have already been performed to enhance the existing SN security features to provide fraud detection. Changes to the method of SNAC charge record generation and an integrated intelligent trend analysis sub-system could form the basis of a fully functional test bed for fraud detection and analysis.

In its current state, the service node fulfils many of the objectives of the IN concept. A service node encapsulates the level of flexibility and adaptability required to minimize the lead time to market of new and innovative services. This study has provided the ability to prototype and field trial IN-based services and concepts in advance of the deployment of core network IN facilities. Services have been tested at low risk using low traffic volumes which, where successful, can be migrated to the core IN for national roll-out.

REFERENCES

1. Hollywood S: 'SCP development in a multiprocessor UNIX environment', International Computer and Communication Conference, Tampa (1992).

2. Sage C J: 'An application programming interface for the intelligent network', XIIIth International Switching Symposium, Stockholm, Sweden (1990).

7

OPERATIONAL SUPPORT SYSTEMS FOR AN INTELLIGENT NETWORK

P E Holmes and N J Street

7.1 INTRODUCTION

Most of the chapters in this book concentrate upon the features that network systems provide and how the network provides them. However, in the world of the telco, this is only part of the story. To offer a usable service to customers, it is also necessary to provide management systems, so that services can be ordered and billed for, that faults and queries can be dealt with and that the performance of the service can be monitored and reported upon. This is the job of operational support systems (OSSs).

It is usual to break down the functions of an OSS into a number of key building blocks, in the form of an architecture. In a functional (or logical) architecture the functions provided by OSSs are broken up into a number of blocks. Each of these has a limited range of functions that are closely associated, particularly with respect to the data upon which those functions act. There are a number of ways of performing this partitioning, but most split the functions into two large blocks — network and service management.

7.1.1 Service management

Service management is principally concerned with how customers see the service being offered. In a reduced form, it can be thought of as how service is provided, maintained, monitored and billed (PMMB being a useful abbreviation to remember):

- provisioning deals with the ordering process with the customer;

- maintenance is all about keeping the service working and dealing with problems;

- monitoring deals with performance reporting;

- billing speaks for itself.

7.1.2   Network management

If service management is about how to manage the service, and therefore the interface to the customer, network management is concerned with the interface to the network. It is principally concerned with issues such as network faults, configuration of network equipment (such as switches, multiplexers), network planning, etc.

In the same way that each new service needs all the functions of service management to be considered, each network needs all the functions of network management to be considered. Note that the two need not be the same — to provide one service can require many networks, and one network can offer many services. For example a future IN network would be expected to offer a huge multiplicity of services.

7.1.3   Why architectures?

A logical architecture shows what must be done in order to manage services and networks. This clear knowledge of the range of management functions provides a baseline to assess the impact of new service types and network technology. An architecture will be a key feature of new OSS designs to meet the challenge of services based on intelligent networks.

7.2   THE CHALLENGE OF INTELLIGENT NETWORKS

The intelligent network concepts developed by Bellcore [1] successfully separated service and feature control from the physical aspects of the network. Service logic embedded in attached processors (for example, service control points) permits changing the flow of a basic telephone call to provide enhanced communications services.

However, this separation increases the complexity of management. Connectivity is established by changing relationships between elements of hardware, software and the links between them. This involves rules, conventions and processes for establishment and termination of such relationships. Failure modes are also more complex, and points of service failure are no longer fixed and are not easy to test in the operational network.

For example, consider a network-based automated call-distribution feature. This feature routes calls to a number of sites on time of day, call load and country of originating call. How does the telco test the service in the event of reported faults? Which of a number of networks are at fault? Is it a physical or a logical fault? Is it the service logic routeing the calls to the network database? Knowledge of previous call load is necessary to determine the correct routeing of any particular call. Will it be acceptable for a customer of a telco who reports a fault on one of its sites, where the access service is provided by telco X, to be told by telco A that its network is alright and that telco X should be contacted?

IN-based services will therefore require management of new physical network elements (the attached processors) and complex configuration of service logic and databases. End-to-end management of communications services across multiple networks will also be required if truly seamless services are to be offered.

However, locating and fixing faults in an intelligent network is not the only new demand on the management systems. For example, the following are key features of IN-based services, all of which create new problems for management systems:

- rapid introduction of new services;

- open-ended range of services;

- volatility of customer and service data;

- increased customer control;

- enhanced signalling systems;

- third-party service provision;

- distributed customer and network data;

- new network elements.

7.2.1   Rapid service introduction

A key driver for the introduction of intelligent networks is their ability to allow operators to design and launch new services quickly, in order to meet the increasing pace of the changes in the needs of the market-place. Within an IN context, this process is referred to as service creation and more detail of the process can be found in Chapter 8.

However, simply designing and launching the service on the network leads to a service for which problems cannot be managed, orders cannot be taken and bills cannot be produced. Clearly, the network enhancement must be accompanied by enhancements to management systems in the same time-scales. Historically, management systems have taken in the order of a year to enhance, whereas intelligent networks are looking at new service launch times in the order of days. Therefore the ways services are currently managed will need substantial changes if these two differing time-scales are to be brought together [2].

If rapid service introduction is to succeed, business processes will also need to change. At present, the processes presuppose a long product-launch cycle — much work will be needed in this area to improve response times.

7.2.2   Open-ended range of services

Currently, the range of services offered on non-IN network platforms is well bounded and understood. However, intelligent networks allow the possibility of a vast number of new and innovative services being created quickly. This can lead to an uncontrolled range of services being available. This causes a number of problems, principally concerned with feature interworking, but increasingly in the control of the range of services offered.

Furthermore, next generation services will not generally be provided individually as low-level features. They will be combined increasingly in 'packages'. These 'packages' can be a mixture of some customer premises equipment and one or more sophisticated network services.

For example, consider Chargecard access to a global corporate virtual private network. To be effective, management must span the entire system across local, national and international boundaries and across the numerous vendor systems likely to be deployed. This will present new challenges to the designers of management systems. They will need the flexibility to manage services in a number of different combinations and still provide compatible operations such as fault reporting and correlation.

7.2.3   Volatility of customer and service data

IN services are characterized by the rapidly changing nature of the customer and service data [3]. For example, a number-translation service is designed so that the dial plan can be changed rapidly in the event of, say, a sudden surge of calls. Managing and tracking this changed data provides special problems. Clearly, volatile data is held on the service control point (SCP), but it is likely that it will also need to be held on the management systems. For example, fault finding, billing and customer reporting will all need access to that data. This means that data management strategies are required which allow volatile service data to be maintained, while managing different versions of replicated data.

7.2.4   Increased customer control

The complexity of IN services, and the increasing business dependence upon them, means that customers will increasingly want to control their own network services, for example, to add new extensions to a centrex service or to increase the bandwidth of a data service. This adds substantial complexity to the management systems that telcos will need to provide. Enhanced security will be a key need, to prevent both the network being brought down by uncontrolled or unauthorized changes, and customers meddling with each other's service configuration. The interface to the customer will also need to be much more user-friendly than is customary with management systems, in order to avoid huge training costs.

7.2.5   Enhanced signalling systems

IN's requirement for complex signalling (see Chapter 3), based around CCITT Signalling System No. 7, also has implications for managing the signalling network. Traffic routeing becomes dependent upon signalling and therefore the characteristics of the signalling network change. This means that the signalling network itself will require new and quicker acting management methods, and this creates a need for separate management of the signalling network. The signalling load and integrity requirements also grow.

7.2.6   Third-party service provision

The window of opportunity to launch a service and recoup the investment is shortening. An increasing number of applications may be only appropriate for niche markets. These may not be economical if a telco has to bear the full cost of service development and deployment. Such services may be provided in collaboration with third-party service developers who can develop these services and sell them to network operators. This could enable telcos to meet their customer demands more quickly and at a lower overall cost. In addition non-telecommunications services (e.g. interactive games) may be offered through the public network of the future.

The introduction of intelligent networks is the first use of an architecture in which a third party may gain direct access into the public telecommunications system. This represents a significant evolution from conventional switching architectures and will, in turn, require a corresponding change in thinking about securing the operation of such networks.

The extent of interconnection between telco intelligent networks remains to be determined, with regulatory bodies such as the DTI [4], European Commission and FCC in the USA certain to influence the outcome. Other uncertainties persist in areas such as future EC directives on the rights of individuals to privacy.

A telco could suffer substantial losses if the limit of open access to third parties is not clearly defined within a secure management framework. As well as the effects of fraud and intentional disruption of the network, losses could result from accidental interference with the signalling network or perhaps from unforeseen service interactions.

Both ETSI and the ITU have working parties addressing this issue, and there is work in progress within the Eurescom programme. However the result of these initiatives is not expected for several years. In the meantime, telcos will implement bespoke countermeasures based on an assessment of service and network asset evaluations and the vulnerability of the physical and software infrastructure to recognized threats and risks.

7.2.7   Distributed customer and network data

With a non-IN network, details of a customer's routeing and network routeing data are localized to a single switch. Even if more advanced services than simple call-delivery are employed (for example, call diversion), only the data in the local switch needs changing. However, IN changes all that. If a customer uses a personal mobility service to change his delivery location from London to Sydney, then a customer calling from Sydney should not be routed through to London before being diverted back. This implies that the routeing plan has to be accessible to all locations on the network. This, in turn, means either a distributed database or a vast, single database with access from anywhere in the world. While this is currently possible on the scale needed to support VPNs for large businesses, neither would be currently possible on the scale which would be needed to provide service to a mass market. The former solution is clearly more scalable, but taxes distributed computing technology to its limits; the latter solution would require a huge central processor and has communications requirements that would be difficult to meet.

7.2.8   New network elements

Intelligent networks bring with them new classes of network elements, and new issues of managing them. The key new system is the service control point (SCP), which is a large computer, sourced from IT vendors and usually based upon fault-tolerant technology. It is usually optimized for rapid database access. The difference between SCPs and standard stored program control (SPC) exchanges is in the degree of specialization. The SPC control computer is dedicated to the job of managing the switch. However, an SCP is a commodity component specialized for a particular job. Therefore managing the SCP as a part of a network starts to introduce computer systems management problems to what had been previously a transport network.

7.3   OTHER MARKET DRIVERS

The challenge for OSS design and implementation is not just to manage IN-based networks — the market is also changing in other areas. OSSs must combine the capability to manage IN-based services while providing the following capability.

7.3.1   Outsourcing

Those corporations who wish to continue building and operating their own private networks increasingly will demand manageable network components and network services. Others will want a smooth and safe migration to a point where a telco, or outsourcing company, is running all or part of the customer's network and performing extensions, upgrades, changes and so on for a predictable fee. Some customers, for example, may want to retain operations of the North American segment of their virtual private network but use an outside source to provide the European nodes. The split may be based on other criteria, for example, the time of day. The management of this telecommunications package is likely to be a shared activity between network operator and the customer — in other words a co-operative activity between the two. This sharing will need to be on a tailored basis. However, in general, most customers will want to concentrate on their business and service management aspects while the telco handles the operational issues of managing the network on a day-to-day basis within agreed service levels.

Finally, a customer may want to sub-contract the entire provision and operation of its information network to a managed services operator. The customer will want to dictate different service level agreements with the telco or operator for various network applications.

7.3.2   The market

Society is becoming increasingly dependent upon telecommunications services. A study conducted in the USA by the University of Minnesota [5] estimates that the loss of voice and data transmission capability could completely shut down a bank in two days and a factory in five. Just one hour without telecommunications services could cost an insurance company $20,000, an airline $2.5 million and an investment bank $6 million. Hence very high service levels will be required for high-value applications and much lower (and hence lower cost) service levels for less vital functions with the ability to change these priorities based on business demand.

The customer will want to dictate the locations to be served, the capacity between and service features to be available at those sites, and may want to vary the price/performance mix of the telecommunications services by time of day/ year or geography. They will demand management information to improve their decision-making processes.

This type of flexible service needs an electronic form of service level management between the customer and the telco. In order to implement these requests the operator (and the customer if continuing to operate part of the network directly) needs an integrated structure between this service view of the network and the network itself so that control can be exercised with minimum human intervention. This leads to a layered architecture for the management of customer network services and the complex information which flows between the telco and the customer.

7.3.3   Correspondent working

Currently liberalization and deregulation of telecommunications services have not advanced to the point where a single telco is allowed to own or control all the elements required to provide access to customers world-wide. So a telco's global network is currently developed in conjunction with other operators throughout the world (known as correspondents). Within the UK customers will increasingly use multiple network operators for their communications services — hence the international call model, whereby a telco is only responsible for a proportion of the call, will become more common within the UK. Interworking and collaboration with other network operators will have to be established. For example, the traffic management problems of telethons and phone-ins which generate very large numbers of calls directed at a single answering point, will not be completely under each telco's control. Without the visibility and control of the network provided by 'collaborative' network traffic management these events could lead to significant network failures and outages.

The challenges this brings will be similar to those posed by the move to personal mobility services. The ability of a customer to move between the fixed, mobile analogue, GSM (global system for mobile communications) and other more localized mobile networks will mean that the operational issues of such networks must be addressed. Not only must the networks physically interwork, the operational support systems of such networks must also be able to interwork seamlessly, so that service provision, maintenance and billing appear unified for the customer, even though the networks themselves may be owned by different operators.

7.4   ENABLING TECHNOLOGIES

The challenge for OSSs in the future can be summed up as the need to be global, rapidly reconfigurable and providing user-friendly interfaces to customers. At present, there is no answer to that challenge. However, a number of enabling technologies have been identified and are being worked on.

7.4.1   Service packaging

Services built on an IN will rely increasingly upon sets of common building blocks, which can be combined rapidly to provide a new service. For example, by combining a 'time of day' package with an 'origin of call' package, a call answering bureau service can be built, with calls being routed to the nearest office during the day, but sent to a central location off-peak. However, this only configures the network. In order to allow management of this new service to be introduced as quickly as the network is changed, the building blocks will need to have management data added to them. For example, pricing details for the package and performance reports that are associated with it could be included. This enhanced package is called a service feature. It is likely that more than one network building block could be built up into a service feature, to provide a higher level set of standard building blocks. The skill of the product manager and the supporting engineering teams would then be to combine these standard building blocks in novel ways to meet customers' needs.

Once a set of common service features has been established, together with the management features that are associated with them, it will be possible to launch the management of those features on the management systems in the same way as the network features are launched on the IN.

7.4.2   Computing technologies

The biggest trends in computing are currently towards putting significant computing power on users' desks, increasing use of client/server architectures for database access, the drive towards distributed computing environments and the cheap provision of high-bandwidth links.

This will lead to systems of increasing functionality and user-friendliness, but at a cost in terms of systems design complexity. Configuration management of large networks of personal computers is acknowledged to present significant problems, but where these systems are critical to managing the company's key services and networks, the need to solve these problems becomes essential. Migration strategies from current systems also need serious consideration.

7.4.3   Standards

The next generation of administration systems are being developed and procured within an architectural framework. It is in no customer's interests to be locked into one manufacturer's products and services where these cannot work with those of other suppliers. Today, the demand is for interoperable components with suppliers competing on quality and support, not proprietary architecture.

Telcos have responded to this demand from customers (and their own networks) with the development of a co-operative approach to architectural specifications for all aspects of communications systems. This approach recognizes the need for partnership, working alongside other vendors and customers to define truly interoperable systems that cope with today's multi-vendor, multi-location networking market-place. For example, BT's co-operative networking architecture takes a step beyond open systems by putting in place the specifications and agreements to ensure that real products and services will be able to work together and conform to national implementations of international standards. It is not static, but evolves with time to reflect changing business pressures and advances in technology.

As one of the largest network operators in the world, BT is investing heavily in systems built against co-operative networking architecture for its own networks, and is now actively implementing these principles into its customer products and services. By harmonizing the architecture and management capabilities of its products and services, BT will offer the customer a true end-to-end managed communications capability within its own portfolio and within multi-vendor networks.

This approach provides a durable, flexible and economical base management platform capability and communications architecture, firmly based on accepted industry standards and value added components. This allows its management system designers to focus on developing those items in the applications portfolio which will reduce its cost base and differentiate its products and services.

7.4.4   Process re-engineering

The consolidation of network administration into centralized operational units still requires extensive manual effort using systems that are not integrated. Future systems will be characterized by automatic flow-through whereby an event, be it a customer order or fault in the network, triggers off a set of actions in an integrated set of computer systems. These then automatically perform the relevant workstring to respond to that event. The processing of the workstring is such that the response has the minimal amount of human intervention necessary.

7.4.5   Advanced information processing

As telecommunications networks become more complex, supporting a much larger variety of services, and management support systems grow in complexity, some of the functions of these systems can only be carried out by employing advanced information processing (AIP) [6-8] to make software systems more adaptable and intelligent. It is a combination of the fields of artificial intelligence and distributed information processing that provide many of the techniques that are necessary in developing AIP solutions to network management. Progress is such that these technologies can be expected to improve the type of software used in most operational support systems over the next five years.

7.5   VISION OF THE FUTURE

The vision of the future of the provision of IN-based services is already well established in many people's minds. It centres on the view of a product manager or marketing manager seeing a niche in the market for a new communications service one morning and then designing that service on a service creation environment on the same afternoon, before launching it on the network the next morning! However, how the management systems would cope with this is less well thought about. The vision that must accompany the one above is of service centres being able to take the first customer's order on the morning of the launch, of the first bill being sent out that evening, all based upon advertising and mailshots that were automatically despatched to likely customers for the service early on the morning of the launch. Clearly, network integrity will restrict the range of services that could be launched in this way, without significant engineering support. However, once that vision is possible, we can claim to truly have a rapid product launch ability.

REFERENCES

1.   Bellcore, SR-NPL-001555, AIN Release 1 Baseline Architecture (March 1990).

2.   Sinoyannis S: 'Service element-based operations requirements', IEEE Proc, 82 (1994).

3.   Robruck R: 'The intelligent network — changing the face of telecommunications', IEEE Proc, 79 (January 1991).

4.   Department of Trade and Industry: 'Consultation document on intelligent networks' and 'Intelligent networks — a response', (1992 and 1993).

5.   AT&T: 'Trends in telecommunications', 7, No 2, p 23 (October 1991).

6.   Helleur R J (Ed): 'Network management', BT Technol J, 9, No 3 (July 1991).

7.   Flavin P G and Totton K A E (Eds): 'Computer aided decision support in telecommunications', Chapman & Hall (1996).

8.   Smith R, Azarmi N and Crabtree I B (Eds): 'Advanced information processing techniques for resource scheduling and planning', BT Technol J, 13, No 1 (January 1995).

8

SERVICE CREATION

G D Turner

8.1   INTRODUCTION

Most diagrams of an intelligent network (IN), for instance the one in Fig. 8.1, show a service creation environment (SCE) as one of the major components of an IN [1]. However, while an SCE is a necessary and key component, it is not sufficient in its own right and there needs to be many other features present to really attain the objective of rapid service creation. This chapter will examine several such features and explain how each contributes to the goal of faster service delivery. The overall process for rapid service creation is then presented and from this it becomes clear that SCE is a hierarchy of tools rather than a single environment, with each toolset being targeted at a particular class of user. These tools are then positioned within the BT service creation architecture. Finally, the chapter will present the vision for total service creation, which takes the concepts of IN service creation but applies them to all of BT's operational support systems (OSS) (see Chapter 7). These systems provide functions such as managing the underlying network, allowing orders to be placed and tracked, faults to be reported and monitored, and providing the ability to charge and bill for new services.

8.2   INTELLIGENT NETWORK SERVICE CREATION

This section examines several of the attributes that are required in an IN to accelerate the creation of new services. They are:

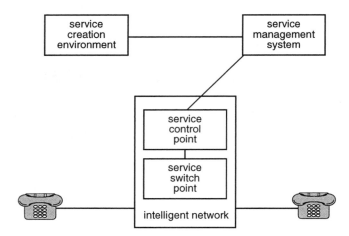

Fig. 8.1   Simple IN architecture.

- functional separation — this separates the basic core functions of real time call switching from the customer and service-specific aspects, so that the latter can be changed more easily;

- portable software environment — enabling services to be developed once and then run on different vendor SCPs;

- generic building blocks — building services from common modules to allow considerable reuse and hence speed of development;

- service logic programs (SLPs) — a simple language for specifying the linkages between building blocks, usually produced by service creation tools;

- graphical creation tools — these tools enable services to be created rapidly by on-screen manipulation of icons which represent the generic building blocks;

- service and network simulators — when a service has been created, it can immediately be simulated on a range of tools that check its functionality, performance, cost, etc;

- on-line deployment — when a service is ready for deployment it can be electronically sent to the network and the appropriate management systems, from the service creation tool.

8.2.1 Functional separation

This is the essential concept of the intelligent network, creating a network intelligence layer of SCPs, separate from the basic switch layer of SSPs. The latter have severe real-time and performance constraints and because there are so many SSPs in the network they need to be highly optimized. Therefore the aim here is to produce a very stable environment that can be exhaustively tested and refined. SSPs need store only relatively small amounts of service or customer-specific data.

In contrast, the SCP is designed specifically to run a wide range of network services. It is also designed to handle large amounts of volatile customer data.

8.2.2 Portable software environment

Technical developments and commercial considerations will invariably lead to changes in SCP hardware. If all existing applications had to be rewritten for each new platform, this would clearly limit the growth of new services. Therefore the approach taken is to produce the applications for a portable software environment that totally isolates them from the physical network and hardware implementation. This approach has similar advantages to using a common operating system across a range of machines.

8.2.3 Generic building blocks

Together with the service creation tools themselves, the use of reusable generic building blocks is the real key to rapid service creation. In the general world of computing, much work has been done on trying to achieve significant software reuse because the benefits are enormous. For example, reusing a program three times effectively triples productivity and, furthermore, on the second and subsequent time it is used, the lead time is virtually zero.

Despite these huge benefits, general software reuse is still rare because it is difficult to realize. However, within the fairly closed domain of the SCP, where the discrete operations of the underlying network are well understood, it is very possible to build generic pieces of software to drive these underlying operations. These building blocks are then called upon in many different sequences to provide the diverse range of IN services.

A typical building block might be 'time-of-day routeing'. This building block would check the user's profile, which may state that after 6:00 pm calls to this number are diverted to a night-watchman. The building block would then check the current time and route the call appropriately.

8.2.4   Service logic programs

Services are created from the generic building blocks by specifying the sequences of these blocks and any conditional linkages between the blocks. This specification is frequently known as a script or service logic program and is usually generated by a service creation tool.

When this SLP is deployed into the network it needs to be 'executed'. The module for doing this is often called a service logic interpreter (SLI) because many IN implementations use an interpreted language [2] for their SLPs. Within BT the term service engine is preferred to SLI, as the flexibility is needed to support interpreted, compiled [2] and possibly other forms of SLP.

8.2.5   Graphical creation tools

Service creation environments (SCEs) or tools represent the user interface into service creation. At present most attention is focused on graphical tools, although in the future other technologies such as expert systems may be used.

The basic concept of a graphical SCE is that icons are used to represent the generic building blocks described in section 8.2.3. The human service creator then drags and drops these icons on the screen and sets up linkages between the icons to represent the flow of the service. These links or flows can also be made conditional upon certain events occurring, e.g. upon a certain number being dialled. This on-screen representation of the service can then be converted into a script or SLP for deployment down to the network SCP. The on-screen display and underlying representation of these tools varies from vendor to vendor. The BT tool set uses the concept, though not the notation, of Fenton and Whitty [3] flowgraphs which are based on graph theory. The Fenton-Whitty theory is also explained and extended in some of the Esprit II Cosmos project documentation [4, 5].

In addition to a service editor, there are usually many other functional modules present in an SCE. Some of these are support functions to do with security access and configuration control, etc. Some are simulators and modules to control the release of services, and these are covered in the next two sections. Other modules may be concerned with particular resources in the network. For instance, there are specialized modules to handle the recording of speech messages that may be played by an intelligent peripheral during a voice dialogue with a customer.

So far this section has assumed there is a single SCE, albeit with some modules being optional. In practice, the requirements of different users are such that no single tool would be adequate or appropriate. Therefore, BT has developed an architecture that identifies three distinct types or levels of service creation tools.

The rationale for these levels is presented in section 8.3 and they are described in section 8.4.

8.2.6   Service and network simulators

Once a new service is created, it needs to be 'tested' and its impact on the underlying network and other services need to be assessed. A range of tools or simulators can be provided in an SCE to automate and support these processes.
   Some of the functions available are:

- functional simulation — this provides a complete emulation of the service, with all the correct look and feel, through some customer premises equipment (CPE) attached to the simulator;

- service logic animation — this enables the service logic to be 'executed' within the simulator with the various icons and links on the screen being highlighted as they are activated;

- network animation — a graphical representation of the network can be displayed, with icons of phones, etc, in various locations, phones at various sites then being stimulated and the resultant traffic flows displayed;

- network performance simulation and analysis — using the same underlying model of the network, conventional traffic theory algorithms can be used to look for bottlenecks and estimated network delays;

- cost and revenue analysis — these tools use estimates of network resources utilization to either cost a new service or make revenue predictions for the service;

- feature interworking analysis — feature interworking is a big problem for IN and not all of it can be prevented in an SCE; however, through various analyses it is possible to detect and reject some erroneous interactions at this stage.

8.2.7   On-line deployment

Once the service is created and tested it is ready to be deployed. The bulk of the software for a service is already present in the network in the form of generic building blocks. Therefore all that requires to be developed is the relatively small service logic program, which can easily be sent electronically.
   In theory a new service could instantly be deployed on to an entire network. In practice it is usual to initially deploy the service into a stand-alone node or

small sub-network so that the service can be tested further and trials undertaken before major roll-out.

8.3   SERVICE CREATION PROCESS

This section describes how these individual IN features can be harnessed together to produce a rapid service creation process. Firstly, it is useful to consider the traditional life cycle for putting a new service into a network, which is:

- design and enhance the physical network — this prepares the underlying network to support the new service, items such as new types of intelligent peripheral perhaps being required for addition to the network (alternatively new switch functionality or capacity may be required);

- develop the basic software for the new service;

- establish the packaging and marketing aspects of the service (How are various combinations of the service to be sold? What is the pricing structure? What are the service level agreements and the arrangements for ordering?);

- if necessary and possible, tailor the service to a particular customer.

Each of these activities can be automated to some extent but the real key to rapid service creation is to do the early steps in advance of the service being identified or required. Each of these can be pre-provisioned as described below.

- Enhance the physical network

   Providing a rich functionality IN means that many new services can be created without requiring new network upgrades. Of course, this is the main justification for an IN.

- Develop the basic software

   This is where the concept of building blocks comes in. Faced with the problem of building a plastic doll's house but without yet knowing the exact requirement, one could produce a set of Lego™ bricks from which any doll's house could be built, once the specification is known. If chosen appropriately, these same bricks could also be used to build other toys such as boats, which are where the real benefits of the building block approach lie.

- Establish the packaging and marketing aspects of the service

  Remaining with the Lego™ analogy. Not only can the bricks be made in advance, they can often be packaged and priced in advance, allowing the customer to specify their own toy from a standard kit. For instance, a certain package may be sold as a car-making kit and would contain some standard bricks and a few 'special' building blocks such as wheels.

- If necessary and possible, tailor the service to a particular customer

  If all the above has happened, then it should be possible for a customer, or a marketing person acting on behalf of a set of customers, to create a service very quickly. Even this process could be accelerated by pre-linking some of the blocks into an example service. This would be the Lego™ equivalent of selling a package with most of the bricks already joined together to form a basic car.

The first item is really network design rather than service creation, but is listed here for completeness. The other items are known as the three levels of service creation — levels 1, 2 and 3.

So far this discussion has centred on making functionality available to the next level in the process and this is shown on the right-hand side of Fig. 8.2. However, in some cases the 'package' available at level 3 will not be adequate to build the service required. In this case, action is required at level 2 to provide the missing capability. This may mean re-packaging an existing piece of functionality or it may require a new building block to be developed at level 1. This process is shown on the left-hand side of Fig. 8.2.

In the worst case one has to cascade all the way down the left-hand side of the diagram, in which case the process defaults to the traditional development cycle described at the start of this section.

8.4 SERVICE CREATION ARCHITECTURE

BT has developed an architecture for service creation that reflects the service creation process just presented. This architecture is shown in Fig. 8.3. Space here does not permit an examination of the service and network management systems. However, for the purposes of this discussion, they are responsible for the storage and distribution of the various outputs from the SCE tools and their deployment into the network.

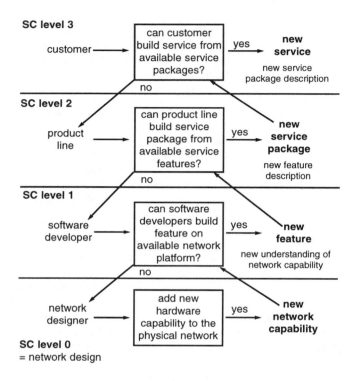

Fig. 8.2   Service creation process.

Within this architecture there are specialized SCE tools for each service creation level which are now examined in more detail.

8.4.1   SCE level 1

These tools are used by skilled software developers to produce the generic building blocks. They are generally based around conventional computer-aided software engineering (CASE) tools [6]. However, they need particularly good configuration management controls.

In addition to downloading the actual building blocks to the target systems, they must also pass a representation of the building block to the Level 2 tool. This representation covers:

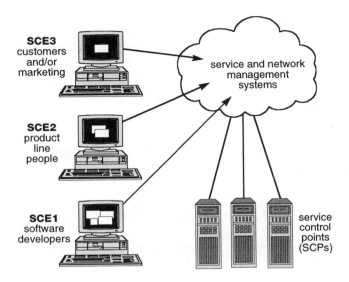

Fig. 8.3   Service creation IN architecture.

- a textual description of the block, for information purposes and as help text;

- its on-screen presentation, i.e. the icon that schematically represents the function;

- the rules governing interactions with other building blocks;

- the data associated with the building block — these will eventually need to be provisioned for a particular customer.

8.4.2   SCE level 2

These tools are used by product line developers or agents acting on their behalf.

From a purely IN service perspective, an SCE2 tool must provide the following functions:

- assemble a basic service, by dragging and dropping icons on a screen and linking them as described in section 8.2.5;

- define how much of the service flow can be altered at SCE level 3, i.e. by a customer — the rest of the service is then effectively fixed for all customers;

- preset some of the data associated with the service with default values, and again define which of these items a customer will be allowed to overwrite.

With the advent of total service creation, as described in the next section, it is now possible and appropriate that SCE2 should do more. In particular, it is the obvious place to define the packaging and marketing information, described as service creation level 2 activities in section 8.3. Examples of these additional functions are:

- define how orders may be placed and tracked;

- define how the service may be charged and billed;

- define how network and customer-reported faults may be handled;

- define what SLAs and performance measurements are allowed.

8.4.3   SCE level 3

These tools are used by customers directly, by BT operators acting on behalf of specific customers or by marketing people acting on behalf of groups of customers. To support these various scenarios, the SCE3 functionality can be provided on a range of hardware, from personal computers (PCs) to sophisticated UNIX™ workstations.

The functions supported through an SCE3 are:

- tailor the service — within the constraints set at SCE2, the user may be allowed to change the flow or structure of their particular version of the service;

- provision some of the service data with customer-specific values.

8.5   TOTAL SERVICE CREATION

In order to launch a new BT service it is necessary to ensure that:

- the network can provide the service;

- there are operational support systems in place that can manage the new service.

The problems of deploying new services in the existing network are well known. All switch vendors need to agree on the new service and build it into their development plans. BT then needs to deploy it, possibly into all exchanges in the network. As a result many new services are deployed on overlay networks. However, this brings problems of network management and interworking, and loss of economies of scale.

Even when an existing network can be used for a new service, there are similar problems at the service management level — for, inevitably, the existing management systems will need modification to support the new service. However, these new requirements are not specified in any standard format and are often not rigorous. This wastes time and effort in interpreting these new requirements. Also many of the current systems are difficult to change and have a backlog of other requirements to satisfy. The net effect is that changes often cannot be made quickly enough and this leads to new management systems being introduced as interim solutions. This effectively creates an 'overlay management system', which has similar problems to an overlay network, namely lack of integration and the sheer cost and logistic problems of supporting so many diverse systems.

Intelligent networks are generally seen as the solution to the first problem, with IN service creation offering the prospect of dramatically shorter time-scales for putting a new service into the network. However, an IN does not help produce the corresponding operational support functions required to manage each new IN service.

Total service creation[1] proposes applying many of the concepts described in section 8.2 to the operational support system domain. Some of these concepts are now being employed in some of the new management system developments but not in any structured way. In particular, the use of graphical tools and automatic deployment is rare. Total service creation provides a framework for producing operational support functionality in a systematic way. The concept of total service creation is shown schematically in Fig. 8.4.

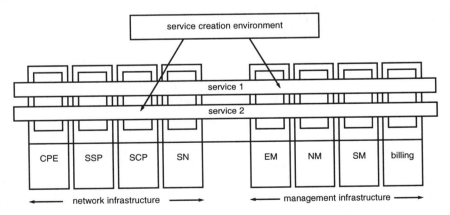

Fig. 8.4   Total service creation concept.

---

[1] Formerly termed 'global service creation'.

8.6   CONCLUSIONS

Total service creation takes the concepts that allow rapid network service creation in an intelligent network and generalizes them to propose 'rapid management service creation' in an 'intelligent management domain'.

When fully implemented, total service creation will enable a service designer to sit at a service creation environment workstation and be taken through a process that defines all aspects of the service and its operational support. This new service could then be simulated and tried on a reference model or a subset of the main network. Finally, given the necessary authority, this workstation could deploy this new service into the intelligent network and simultaneously upgrade all the appropriate management and billing systems to support this new service.

Adopting the approach of total service creation will enforce a common method of defining new services and the associated management features.

Total service creation will enable many services to be provided in a few weeks or months rather than years. This will mean large savings in terms of the costs and number of people involved in developing the current range of operational systems. However, teams will still be required to develop and continuously enhance the systems that support total service creation.

REFERENCES

1.   'The Intelligent Network', Northern Telecom, Communications International, EMAP Business Publishing Ltd (1992).

2.   Krutz R L: 'Microprocessors and logic design', John Wiley & Sons (1980).

3.   Fenton N E and Whitty R W: 'Axiomatic approach to software metrication through program decomposition', The Computer Journal, 29, No 4 (1986).

4.   'Cosmos (Esprit II Project 2686)', Technical Annex (December 1988).

5.   Pengelly A D and Fuchs N: 'Software structure and cost management — the Esprit II Cosmos project', BT Technol J, 9, No 2, pp 39-46 (April 1991).

6.   Fisher A S: 'CASE — using software development tools', John Wiley & Sons (1991).

9

TOWARDS KNOWLEDGE-BASED NETWORKS

P Willis and I G Dufour

9.1 INTRODUCTION

The convergence of computers and communications is producing many changes in society. The technology driving these changes will be a network of machines processing information. This network will appear to possess knowledge since it will be delivering the correct information to the correct people using both knowledge of the information and knowledge of the people and hence it might be described as a knowledge-based network.

The path to this future vision is influenced by three technological mega-trends summarized as bandwidth, intelligence and information, although, as will be shown, these terms need some qualification. To set them in context it is first necessary to review the pre-convergence history of telecommunications and computing technologies.

9.2 PRE-CONVERGENCE HISTORY — TELECOMMUNICA-TIONS, LAN TECHNOLOGIES AND COMPUTING

The key trends in telecommunications developments over the last two or three decades have been the use of computers to control switches, the dominance of digital techniques, the availability of huge bandwidths via optical fibres in the fixed network and an increasing use of cellular radio networks which themselves are highly dependent on computer control. Asynchronous transfer mode (ATM) switches are starting to appear, bringing the traditionally speech-dominated telco world and the datacommunications world together and providing an infrastructure for broadband multimedia services.

Key trends in local area network (LAN) technologies are increasing bandwidth, increasing efforts to interconnect LANs — more businesses are using enterprise networks which are formed from the interconnection of many LANs — and an increasing use of standards to allow for multivendor environments. The term 'information network' was often applied to these enterprise networks because information technology was applied to office automation, but the term is now taking on a very different meaning as will be discussed later.

In computing, the key trends have been ever-increasing speeds of operation, the development of large, reliable, centralized, mainframe computers, the development of the personal computer (PC) and UNIX-based workstations, and most recently the development of client/server architectures allowing intelligent terminals (clients) to work with network-based servers. Now further developments are being seen where processing power and data can be widely distributed — a concept that maps particularly well with distributed intelligence in networks (see Chapter 13). The impact of microprocessors on customer premises equipment (CPE) is creating a vast range of intelligent appliances that can be connected to, and can co-operate with, the core telecommunications network, introducing the concept of co-operative intelligence.

9.3 BANDWIDTH, INTELLIGENCE AND INFORMATION

Bandwidth, intelligence and information are three major areas influencing the development of knowledge-based networks and each is reviewed in the following sections.

9.3.1 Bandwidth

Bandwidth is an analogue term traditionally used to indicate the number of hertz required to carry a particular signal. It is ultimately an expression of information-handling capacity. That is why it is also now used as a synonym for data rates in digital systems. High bit-rate systems are said to be of high bandwidth. Bandwidth is used here in its widest sense, as are its counterparts narrowband and broadband.

A key development affecting bandwidth trends has been the invention and relatively rapid deployment of optical fibres. The cost per bit transported has reduced significantly and yet currently only a fraction of the optical fibre's capacity is accessible [1]. Another key trend has been the ability to exploit copper wires for greater bandwidth [2]. At the same time compression technologies allow high bit-rate source material to make efficient use of available bandwidth. Bit rates can also be traded with time in conjunction with digital storage capability. A film can be sent over a narrowband medium overnight, or over a broadband link in seconds — in each case the material is stored at the receiving end for viewing as required.

9.3.2 Intelligence

The most conventional telecommunications view of an intelligent network (IN) as defined by ITU-T is where service control is separated from the network routeing logic (see Chapter 1). Another view is that it is simply about the convergence of telecommunications and computing. A further view of an intelligent network is of a humanized network embodying knowledge about the calling and the called parties and providing appropriate voice responses to voice inputs. The simplest way to imagine this is by reference to the manual switchboard operator of a hundred years ago, i.e. the intelligence, knowledge and information held by the operator is replaced by data and logic embedded in computer programmes. These concepts can be extended to embrace non-speech and broadband networks. This view demonstrates that the use of the word intelligence in IN is something of a misnomer. Nevertheless, machine intelligence, as postulated by Alan Turing as long ago as 1950, is now becoming a probability rather than a distant possibility. It can be seen, therefore, that there is considerable ambiguity over what is meant by intelligence.

Today, the services that a so-called intelligent network enables are those such as number translation services (FreeFone, Premium Rate), virtual private networks (VPNs), cashless calling (Chargecard) and cellular radio services. In the medium term these are being modified for a greater degree of tailoring to meet specific customer requirements. Examples of this are time-of-day routeing variations (advanced number translation services) and call delivery determined by caller location (e.g. a national store chain directing calls to the branch nearest to the caller). In addition, capabilities such as number portability and personal numbering will appear, as will non-speech services and eventually broadband and bandwidth-on-demand services. At this stage the network is smart, but it is unintelligent in human terms. It is a processing machine with flexible support systems and with a telecommunications interface to the customer. The computing capability is distributed and the network comprehends total mobility.

Beyond that, the telecommunications network may have gained enough intelligence through service control computing power to move on to the next stage. Here, core-network intelligence will work co-operatively with ever more intelligent terminal equipment, where both use machine intelligence to respond to the user as an individual not only taking account of location and personal profiles, but also comprehending context, emotion and other human characteristics. The network will thus be able to determine its actions in response to a wide range of inputs. It is at this point that the network can be better described as a knowledge-based network. Human to network communication for control purposes will almost certainly be dominated by speech even for controlling non-speech services. It will be in an environment where numerous service providers produce information or entertainment content ranging from digital libraries to video databases.

9.3.3   Information

For some time we have heard that we are now in the information age. This is usually taken to mean that we are passing from the industrial society to the post-industrial society. Put another way it means that service industries are becoming more important in some ways, not least in numbers of people employed, than manufacturing industries. In this context, those employed in 'service industries' certainly include people working in hamburger shops, but it also means teachers, lawyers, management consultants, computer programmers and endless other occupations where people generate or use information for their work. Some, such as Peters [3], take the view that the 'service' content even in manufacturing industries is high. What matters is that a growing band of information-intensive workers need to access and exchange information via telecommunications and the infrastructure to realize this will revolutionize telecommunications itself. The network to support this requirement is increasingly known as an information network, or as the 'information superhighway', which embraces the traditional telecommunications network, the terminal appliances and invariably the information content as well.

A view of the impact of this revolution in telecommunications can be found in Naisbitt [4]. In this he defines telecommunications as encompassing telephones, televisions, computers and consumer electronics and regards it as the driving force that is simultaneously creating the huge global economy and making its parts smaller and more powerful. He notes that: 'In the global economic network of the 21st century, information technology will drive change just as surely as manufacturing drove change in the industrial era.' This is taken a stage further by Negroponte [5] who distinguishes between an existing world order based on atoms (i.e. tangible products) and a new emerging one based on bits (i.e. information and entertainment in digital form).

9.4   KNOWLEDGE

The use of the word 'information' in the above section can frequently be substituted with the word 'knowledge'. While definitions of these terms are notoriously difficult, for convenience knowledge is assumed here to imply information with meaning. We are moving towards knowledge workers dominating a knowledge-based economy, both of which are highly dependent on telecommunications. The social and political implications of this movement were set out some years ago by Toffler [8]. His book includes a chapter on 'extra-intelligence' where '......we are looking towards networks.......(which).........act on the outside world, adding extra-intelligence to the messages flowing through them' and a chapter on 'net power' that notes that '......in the super-symbolic economy, it is knowledge about knowledge that counts most.' The latter point has important implications for the design of networks if they are to support knowledge-based workers. Negroponte [5] takes this aspect substantially further forward and extends the principle to the general day-to-day requirements of individuals and the impact machine intelligence can have on them and their surroundings.

9.5   THE BIRTH OF KNOWLEDGE-BASED NETWORKS

The problem with identifying intelligence in an intelligent network is, as has been shown, due to the ambiguity of the term intelligence. It is more reasonable to identify where knowledge is located. A knowledge-based network is easier to understand in terms of what it does and what it offers the consumer. The consumer can buy intelligence in the form of processing power and software at lower cost than a telco can sell it in the form of network intelligence. However, the consumer may look more favourably on paying for knowledge obtained from a knowledge-based network because consumers understand that they cannot buy this knowledge cheaply. Today's intelligent networks have limited knowledge, mainly knowledge of numbers, time of day and location, rather than being a knowledge-based machine using inferencing on the raw information.

9.5.1   The technology of knowledge-based networks

The convergence of telecommunications and computers to form knowledge-based networks is happening from several directions:

- computers are being used widely within telecommunications systems, e.g. modern telephone exchanges can be described as specialist computers;

- general-purpose computers are being used for additional control of telecommunications, e.g. intelligent networks (see Chapter 1);

- all communications can now be represented digitally (e.g. voice and video have been transported digitally for many years) — this has been the main driver in the multimedia revolution because all information and entertainment content can be represented as a stream of bits;

- ATM is seen as key to providing a single network infrastructure capable of supporting all forms of media;

- computer systems have been using data communications for decades (terminals connected to remote mainframes), but, recently, computers have had interfaces added to them to allow them to control non-data telephony calls directly (see Chapter 10) — from the computer's point of view the telephone has become a peripheral;

- ISDN eliminates the different interfaces to voice and data networks by providing a digital interface;

- the increasing size and interconnection of enterprise networks, and public data networks such as the Internet, means that the data network is increasingly beginning to act like telco networks, i.e. a computer user can communicate with any other computer, or computer user, in much the same way as a telephone user can contact any other telephone user;

- intelligent terminal equipment increasingly looks like a computer, either full-function multimedia or with reduced function (e.g. set-top boxes, lite-PC, games machines), each co-operating with network-based intelligence.

Beyond this there is no consensus between the computer and communications industries over what form the network will take and it is most likely that it will be driven by the equipment that consumers buy. While a broadband infrastructure is required for many new services, the range of services possible over an essentially narrowband infrastructure is still large, partly because bit rate and time can be interchanged for entertainment services, and partly because a knowledge-based network will select and minimize essential information.

The future form of knowledge-based networks will also be influenced by non-technological factors such as:

- the development of standards for applications, peripherals, interfaces and design tools;

- the cost;

- the deployment of intelligent networks and services such as Internet by computer and communications companies;

- regulation.

9.5.2   The 'location of knowledge' debate

The computer, communications, information appliance, media and entertainment industries are currently competing to dominate the converging market-place which will, in practice, be dominated by those who have the most value to add. The most value will be added by those who own and control the knowledge, and the knowledge-based networks, in the converged market-place.

What is a network? The answer to this question will be influenced by whether you are a telecommunications engineer, a computer scientist, a psychologist or a politician. In general, a network is defined as a set of entities and the relationships between them, which have been formed for a co-operative purpose. The numerous views of a network are shown in Fig. 9.1 which bears some resemblance to the ISO OSI 7-layer model. However, the intention of Fig. 9.1 is to demonstrate the views people have of networks and not to present a view of their technical composition.

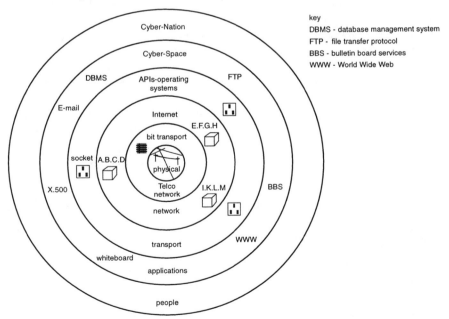

Fig. 9.1   What is a network?

In the centre of Fig. 9.1 is a circle representing the physical world. To the people in the physical world a network is made up of cables, fibres and telephone poles. Such people only think in terms of bandwidth — to them a national information infrastructure is copper or fibre in the ground with no consideration of the switching equipment and CPE required to use the bandwidth. This physical network has its own method of labelling, or addressing, the cables and fibres.

In the next layer are traditional telco people. These telco people take the cable infrastructure for granted and think that the network is formed by a collection of telephone exchanges (or switches). To the computer world the telco network provides end-to-end bit pipes. The telco network uses telephone numbers to address end connections.

At the network layer, or the Internet layer, reside routers and CPE. These have unique network addresses. At this layer the network is formed by a collection of subnets, routers and hosts (end-systems). The network is a logical abstraction of the physical and telco networks (or bit-pipes). Network layer people think in terms of offering logical connectivity between machines. Traditional IN functionality could be considered to reside at the network layer as it deals with the routeing of telephone connections.

At the transport layer are services provided by operating systems, accessed via the APIs (application programming interfaces) offered by the operating systems. This is a software world used by application writers. Variants of APIs exist, some of which are over 20 years old, and programmers tend to use their favourite, a popular one being 'sockets'. To programmers the network looks like various transport services between machines, literally a socket offering a transport service, e.g. connectionless data, to other machines. Programmers need no knowledge of the details of the Internet, the telco network or the physical network.

At the applications layer sit the applications that people use. The applications are written by programmers who use the services of the transport layer. Applications present a wide range of interfaces to users, ranging from applications where the user has to be network-address aware or telephone-number aware, to applications where no knowledge of a target connection is required. In the Internet this application layer is often called Cyber-space. Most people have a limited knowledge of the area of Cyber-space they use and usually no knowledge of the lower network layers is required. The preferred addressing mechanisms in Cyber-space are addresses with which people can associate, for example names of people.

The ultimate network is that formed by the people that are using the information network. The term Cyber-nation has been used by people in the Internet to identify the people that use it. The Cyber-nation is essentially a nation without geography. When politicians discuss the information superhighway or the global information infrastructure they are invariably concerned about the impact of the information network on society (and their own future). Globally available bits threaten the existence of the nation state, but no one yet knows what might challenge it, although it is already postulated that the number of 'countries' may increase [4].

9.6   THE ERA OF KNOWLEDGE-BASED NETWORKS

As has been outlined above, the future will increasingly be dominated by intelligent telecommunications networks and intelligent terminals. These will jointly sustain a wide range of applications involving many information and entertainment types provided by independent service providers. The future should thus bring about increased competition in the areas of information technology and communications. There will be several communications networks using slightly different network technologies forming knowledge-based networks just as roads, rail and air provide a variety of transport systems. There will also be a range of intelligent terminals, such as smart TVs, multimedia phones, multimedia PCs, lite PCs, set-top boxes and even games machines. These intelligent terminals will have more technological similarities than dissimilarities. They will include specialized boxes to offer ease-of-use for very particular information or entertainment retrieval and processing functions. Every home will have several and they may communicate with each other as well as with core networks — they will be information appliances.

Information appliances have been with us for some time. The pocket calculator is an example of an information appliance. It does just one job, but it does it very well and it is easy to use. If any domestic appliance is considered, a cooker, a television, a kettle, they all perform a limited function very well. Appliances are easy to install and use. Plug them into the required services, e.g. gas or electricity, and turn on. General-purpose computers as used for most of today's information processing are not appliances — a general-purpose computer is not a simple-to-use device that performs one task well. In the future computers will still be used but they will be highly specialized, performing a limited range of tasks very well. The information appliance will be a plug-in and turn-on box that will process information in a certain well-defined manner. For example, an information appliance might be a knowledge-retrieval system, it might be the size of a button and use only an audio interface, and could connect into global knowledge-based networks via wireless links.

The suppliers of multimedia services information will become utilities. The information appliances will plug into utilities like today's domestic appliances plug into the gas, communications or electricity utilities. We will depend on information utilities just as we depend on electricity utilities today. The loss of an information utility in the future will have as grave an impact as the loss of an electricity utility today.

The form of communications networks is harder to predict. However, imagine a world with three major types of providers of communications networks:

- interactive cable TV networks;

- multimedia telephony networks;

- packet data networks.

These networks will interconnect for a range of reasons, including customers seeing added value in being able to obtain information from suppliers residing on other networks. Also, businesses will want to communicate with everyone regardless of their network connectivity. There should be no discrimination between consumers and providers — anyone will be able to be a seller of information as well as a consumer, although in practice information providers may have different bandwidth and service requirements from consumers. Interactive cable TV, the multimedia telephone and the multimedia workstation are all versions of very similar information appliances. The main difference in these information appliances is that they are marketed differently. Some information appliances may be capable of obtaining their communications services from different types of communications networks. An information network is not complete without the appliances, or the people that are providing and consuming the information. The various communications networks may interconnect via some backbone networks and these may even be global backbone networks. Some networks will have peer arrangements with other networks and set up their own interconnection. Mobile information appliances will be common. People will use portable PCs, organizers and personal intelligent communicators (see Chapter 11) which will connect into the global network using wireless techniques.

Information consumers and providers, with their information appliances supported by the information utilities, will form the tele-information market-place. Locating where intelligence and knowledge lies within this complex web is not straightforward. Although all the information appliances have similar processing power they are different, for example:

- the multimedia workstation connected to the Internet has considerable intelligence, while the network has small amounts of intelligence (a little knowledge exists in the Internet to provide routeing and naming information);

- the multimedia telephone may be un-intelligent with most of the intelligence provided by the knowledge-based network;

- interactive cable-TV set-top boxes may be somewhere between the intelligence of the workstation and the telephone, with content and knowledge in the service provider's domain.

A knowledge-based network has the possibility of bringing the consumer closer to the knowledge in a more efficient manner than in the Internet. One of several problems with the Internet is finding where the knowledge is located whereas this will not be a problem with a knowledge-based network.

9.7   CONCLUSIONS

The essence of convergence is digital technology — telecommunications, datacommunications, computing, media content and applications are all digital. Communications networks are becoming closely tied to computing capability forming a distributed computing network. The network operators of the future will therefore be operating a huge distributed computing network. This network will process information by applying intelligence to it in the manner understood by machine intelligence rather than that currently understood by IN. The network will appear to its customers to have knowledge, it will appear to provide and operate on knowledge both of the people and of the information, and it will therefore be a knowledge-based network. The knowledge-based network will come to fruition in an era when the economic, social and political structures will be dominated by knowledge and information within a digital environment and in this future converged world there will be many players with constantly changing boundaries between them.

REFERENCES

1.   Cochrane P and Brain M C: 'Future optical fibre transmission technology and networks', IEEE Communications (November 1988).

2.   Cole N G: 'Asymmetric digital subscriber line technology — a basic overview', BT Technol J, 12, No 1, pp 118-115 (January 1994).

3.   Peters T: 'Liberation Management', Macmillan, London (1992).

4.   Naisbitt J: 'Global Paradox', Nicholas Brealey Publishing (1994).

5.   Negroponte N: 'Being digital', Alfred A Knopf, New York (1995).

6.   Toffler A: 'Powershift', Bantam Books (1990).

10

DISTRIBUTED INTELLI-GENCE AND DATA IN PUBLIC AND PRIVATE NETWORKS

R P Swale and D R Chesterman

10.1   INTRODUCTION

In the early days of telecommunications networks — and telephone networks in particular — call completion was dependent upon manual switching and there was a strong reliance upon human intervention to deliver the end service. While this structure afforded the opportunity of a near infinite range of supplementary services, it was also open to less attractive features such as inconsistency in service and feature delivery — not to mention fraudulent misuse. The adoption of electromechanical switching resolved many of these issues. However, while this produced an inherently more consistent network and associated service offering, it did so at the expense of the 'intelligence' employed within the network.

The application of digital computing technology to telecommunications network systems has afforded the opportunity of developing a wider range of services and facilities. These developments may be considered to be in some way replacing the previously lost 'network intelligence' and are reintroducing scope for network operators to offer an increasing range of enhanced services. Indeed, in recent times public network operators have sought to support their customers' needs for using telecommunications as a way of obtaining the 'edge' in an increasingly competitive market-place. This has largely been achieved through network-based services, such as Freefone. A key feature of the realization of these services is their strong reliance upon service-related data and the manipulation of that data in the operation of the service by 'intelligence' functions. This is the essence of present-day network intelligence. While it is possible that for many services all the necessary information could be held and processed by a single modern digital exchange, it is apparent that there are significant operational difficulties with this approach for anything other than the most trivial of network topologies. For example (ignoring the concept of 'overlay networks'), in a reasonably sized network employing a number of switches, each switch in the network would need to retain both the number translation and charging data for each Freefone number operated. Furthermore, any updating of the service data is likely to involve data changes to all network switches which will invariably take time to complete, especially if a number of different switch types are used in the network.

The effects of increased reliance upon service data, the need for manipulation of that data by the network and the distribution of the data throughout the network topology has been the motivation behind the 'intelligent network' (IN) techniques discussed in this book. However, solutions based upon these principles to date have tended to be implemented on a platform per service basis which has largely migrated the issues of service logic consistency and minimization of data duplication away from the switching domain and into the domain of service control platforms realized by computing resources. Furthermore, while the IN approach has emerged as a desirable architecture for the public network in the delivery of advanced services, private telecommunications systems have seen the complementary emergence of computer/telephony integration (CTI) as the route to providing tighter coupling between business information systems and 'front-office' call handling in the quest for greater business efficiency.

This chapter provides an overview of both the public and private domains and presents an assessment of aspects of the computing issues for realizing both public and private intelligent networks. As public IN systems have been discussed elsewhere, this chapter explores private systems and CTI in greater depth.

10.2   COMPUTERS AND TELECOMMUNICATIONS

The network intelligence playing field may be considered to consist of computer-based intelligences linked together by communications capabilities. While these two fields of computation and telecommunications have originated separately, it is apparent that their relationship has grown somewhat unclear over the past 10 - 30 years. For instance, the introduction of common control (computer-based) switching systems within the telephone network and the development of complex communications systems (such as Ethernet) within the computing environment may be seen as examples of this theory in practice. It has also been observed that developments within the computing field have in many ways preceded those in the public telecommunications field — as confirmed by a consideration of computer system architectures [1]. From such material it is apparent that there is a trend towards processing capabilities migrating to the edges of the physical architectures employed. This may be envisioned using a subjective tool termed an 'intelligence distribution graph' as shown in Fig. 10.1 which presents characteristics for centralized host systems, network-interconnected host systems and also the distributed computing systems typified by client/server architectures.

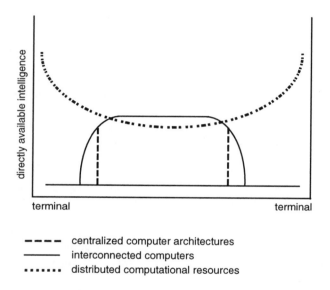

Fig. 10.1   Intelligence distribution graphs for various system architectures.

However, when making such analogies, it is important to recognize that the stringent, non-functional characteristics of telecommunications systems do not tend to be matched by commercial computing platforms. As a consequence, while much may be learned by studying computer system architectures, the solutions proposed may not directly translate into the telecommunications environment. Even so, it is apparent that, as computing platforms evolve, there will be an increased reliance upon and integration with telecommunications facilities to meet the demand for both increased access to data and other resources and the sharing of computational load. The migration towards distributed systems architectures will therefore force alignment between the non-functional characteristics of these two domains.

10.3   DISTRIBUTED INTELLIGENCE AND DATA IN PUBLIC NETWORKS

Within modern public telephone networks, network services are currently delivered either through service logic embedded within switching systems or through the use of the emerging IN-type control architectures. The following discussion presents a consideration of these architectures from the point of view of how service logic and data are distributed.

10.3.1   Non-IN approach

Until the development of IN-type control architectures, network-based intelligent call-handling services were generally implemented as part of the software within the stored program control (SPC) telephone exchanges used to realize the public network. Within the multivendor switching environments operated by most telcos throughout the world, this causes a number of significant problems which are well understood and have been a motivating factor for adopting IN control architectures. These problems are mainly centred on the use of multiple switch types which give rise to variations in service logic implementations and also the reliance upon switch vendors to deliver the service logic as part of exchange software drops. These effects combine to render a consistent overall service offering both difficult to achieve and time consuming to deploy. A further disadvantage of this approach, which results from embedding the service logic within the switching system, is that the service data is also bound to the switching system. In the case where several switches are used this may cause service data to be replicated across several switches. This causes management overheads in administering the service, consumes more exchange resources, such as memory, due to data duplication and creates difficulties in implementing subsequent services. This last point is important when personal mobility services are considered.

10.3.2   IN structured networks

Most major telcos throughout the world have experimented with IN-type control architectures in trying to overcome some of the deficiencies found with switch-based service logic and data. Indeed, IN techniques have proved their worth in providing consistent service offerings through the provision of services independent of the underlying switching systems. However, since few public networks fully support IN control systems from all exchanges, difficulties remain in that customer service logic and data is now distributed across both SPC and IN control platforms. In this sense IN is still far from being the panacea it could be.

This is compounded by the fact that at present IN platforms tend to be established for particular services as they are identified and brought to market, which leads to scope for further duplication of customer service data across several platforms.

10.3.3   Observations

Within the public network, there is a major drive towards migrating service logic away from the core network switching systems in order to provide a more flexible approach to service provision. To avoid a simple migration of the problem from the switching fabric to the IN control platforms, there is the need to maintain a single consistent view of the service logic from two aspects. Firstly, there is the need to ensure that all parts of the switching fabric can access the same service logic and data to support possible services such as personal numbering and personal mobility. Secondly, there is the need to ensure that associated operational support systems (OSS) can obtain a single logical view of operating services to provide a consistent view of services to the various 'front-office' systems. With the potentially enormous volumes of data implied by a full public network, both of these factors imply the adoption of a distributed systems approach to the platform solution adopted.

10.4   DISTRIBUTED INTELLIGENCE AND DATA IN PRIVATE NETWORKS

As with public telecommunications networks, private networks have emerged from a background of manual switching. The drive towards providing enhanced facilities has seen migration towards the use of computer-based systems to deliver the required service logic. However, it is considered that the drivers within the private arena are somewhat different to the public network in that the profusion of facilities has largely exceeded saturation point and the development path now focuses upon applications rather than specific facilities — as will become apparent from the following discussion.

10.4.1   Embedded switch functions

Within private networks and switching systems, there is a parallel with the non-IN-based solutions of public networks where facilities are embedded in the switch platforms. This has led to the proliferation of a wide range of facilities on private switching systems and the use of enhanced private network signalling systems, such as digital private network signalling systems (DPNSS), to allow many facilities to operate throughout a private network. Unfortunately, it is also now becoming widely recognized that many of the enhanced facilities are seldom, if at all, used — largely due to the user-unfriendly interface supported, which is based upon raw dialled digits and the incompatibilities of features and feature invocation from one manufacturer to another. Just as with the public environment, where there has been a drive towards removing service logic from switching systems to allow network services to be more rapidly created and delivered to the market-place, customer premises equipment (CPE) has largely been driven by the ability to provide ever more powerful facilities as a means of product differentiation.

10.4.2   Computer/telephony integration

Since the advent of digital telephony more than a decade ago, the convergence of computing and telecommunications in CPE has been an obvious and much heralded concept. However, discussion which is centred on the physical convergence of the LAN and the PBX has generated little interest among potential users, since they are generally not concerned about the physical infrastructure on which their services are delivered. Instead, it is the convergence of applications, and the appearance of standards to support this, which is fuelling the current rapid growth in computer/telephony integration (CTI).

There are two basic approaches to CTI, referred to as 1st- and 3rd-party CTI. These are discussed here.

10.4.2.1   1st-party CTI

1st-party CTI is characterized by the fact that integration takes place at an individual user's telephony ports, the computer (typically a PC) being given access to the signalling capabilities of that port. Figure 10.2 illustrates a number of possible configurations:

- the first of these is typical of a PBX digital featurephone, in which the phone provides a serial data port which gives access to both the signalling capabilities of the port, and the user (voice) channel for data transfer — while the PC may be used to make, answer and log calls, the handset of the featurephone would normally be used for speech;

- in the second configuration, the PC subsumes the full functionality of the phone;

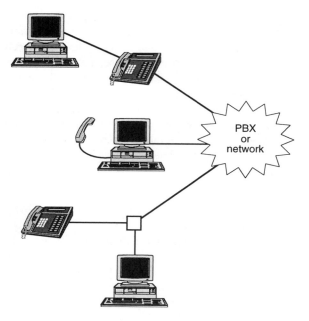

Fig. 10.2   1st-party CTI.

- the final configuration is typically used on analogue extension ports, and uses a 'smart box' to allow the telephony signalling to be presented to the PC's serial data port — the 'smart box' may also include modem functionality for data transmission with speech provided via the phone's handset, but both the phone and the PC may take or share control of a call.

It can be seen that 1st-party CTI either provides the PC with a serial link for telephony control, or requires the addition of a telephony card to the PC. There are a large number of such cards now available, providing analogue (2-wire or 4-wire), ISDN basic rate (144 kbit/s), primary rate (G.703, 2.048 Mbit/s), or asynchronous transfer mode (ATM) interfaces. Some cards have multiple ports and supported signalling systems include L/D, MF, CLASS, ISDN (Q.9xx), ISDN30 (DASS2), DPNSS, and even CCITT SS7. It is also possible to include multiple cards within a PC chassis, and proprietary or *de facto* standards have emerged for the interconnection of speech channels between cards within the PC, typically on 2.048 MBit/s ribbon connectors. Typical of these are SCbus (part of Dialogic's signal computing system architecture — SCSA), and MVIP (multivendor integration protocol). Additional functionality available on PC cards includes modem facilities, fax encoding, voice and video codecs, speech compression, and 64 kbit/s switching.

Applications for 1st-party CTI are almost unlimited, and currently include such diverse facilities as screen-based featurephone, auto-dial from a screen directory, videophone, PC-based answerphone, integrated e-mail and voice-mail, facsimile services, and call-logging. Full multimedia communications via ISDN is available, and is exemplified by BT's VC8000 product.

The availability of new applications is likely to be considerably accelerated by the inclusion of a telephony applications programming interface (TAPI) in Microsoft's Windows 95 [2]. This enhances the normal Windows API with a set of functions which may be called to enable telephony events to be monitored, and telephony commands to be given. Figure 10.3 illustrates the TAPI environment, showing a variety of applications accessing the Windows libraries, with third party 'service providers' giving access to the capabilities of various PC cards and network services.

Figure 10.4 illustrates how voice-mail might be integrated with e-mail in a typical application (in this case, Microsoft's 'Mail'). The voice-mail is distinguished by the phone icon.

There are a number of points to note about 1st-party CTI.

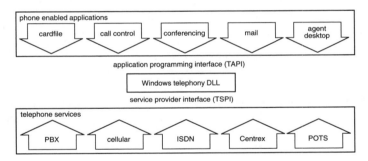

Fig. 10.3   The TAPI architecture.

From	Subject	Received
✉ Swale, Richard	RE: Article for BTTJ	03/01/96 14:33
☎ 0171-492 8800	Telephone Message	04/01/96 08:21
☎ 01473-643210	Telephone Message	04/01/96 10:53
✉ Chesterman, D	Notice of Group Meeting	05/01/96 09:35
☎ 0171-492 8800	Telephone Message	05/01/96 10:29

Fig. 10.4   Integrated voice and e-mail.

- Applications are not limited to the desk top. The availability of primary rate cards makes the provision of centralized or networked facilities a practical reality. Indeed, BT's own audio and videoconference services are provided in this way. The distinction between 1st- and 3rd-party CTI (see below) becomes very blurred in such applications.

- 1st-party CTI relies on normal telephony or ISDN ports, and can therefore be applied to any system (e.g. PBX, centrex, PSTN, etc) without enhancement to the switch.

- A LAN is optimized for bursty, asynchronous, high-bandwidth communications, and therefore provides the optimal solution for the transport of applications data. Equally, the telephone system has been optimized for delay-sensitive, and constant low-bandwidth communication. Thus, in a normal office environment it is normal to provide both media (as illustrated in Fig. 10.2), with the telephone/ISDN networks only being used for data communications where the user is working remotely (e.g. teleworking). However, technologies such as ATM and isochronous ethernet (isoENET) are becoming available in the CPE environment, and these will make integration at the transport level more common in the future.

- The wide variety of interface cards available make it possible to build a complete PBX in a PC, and at least one such system is now being marketed. While the benefits of such an approach for simple telephony are doubtful, the advent of true multimedia PCs combined with communications media such as ATM would seem to point to a future where the LAN hub and PBX become one.

10.4.2.2   3rd-Party CTI

Figure 10.5 illustrates a basic 3rd-party CTI architecture, and it can be seen that this very closely parallels an intelligent network. For this reason, 3rd-party CTI is sometimes referred to as 'private IN'.

These systems rely on the addition to the telephony switch (PBX, ACD, or centrex) of a CTI port, which provides a telephony command and status link to a host computer. The capabilities provided via this link are similar to those available through 1st-Party CTI, these being the ability to monitor activities on the switch ports, and to assume the control of calls through primitives such as 'Make_Call', 'Clear_Call', 'Transfer_Call', etc. However, there are two important additions:

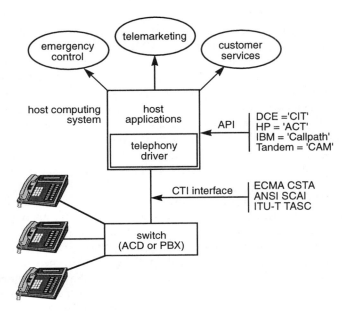

Fig. 10.5   3rd-party CTI.

- it is often possible to take control of the routeing of a call before it has been delivered to a port;

- the host computer system has the opportunity to monitor and control **all** calls and devices on the switch.

3rd-Party CTI messages are carried on a variety of systems, with X.25, Ethernet and (to a lesser extent) ISDN layer 2 being the most common. Most PBX manufacturers have defined their own proprietary message sets, although several standards have now been ratified. These are:

- CSTA (computer-supported telephony applications) — defined by task group 11 of technical committee 32 of ECMA — although this is a European standard, it has won the active support of the majority of major US switch and computer vendors;

- SCAI (switch-controlled applications interface) — defined by ANSI (T1.626), this standard tends to be favoured by those providing virtual private services (e.g. the Regional Bell Operating Companies (RBOCs));

- TASC (telecommunications applications for switches and computers) from the ITU — this work was started in an attempt to bring the work of ECMA and ANSI together into a single standard; although the first issue of TASC has been ratified, work on this standard has now been suspended.

Tables 10.1 and 10.2 compare the capabilities of the CSTA and SCAI protocols, and also include the equivalent INAP primitives. It should be noted that these standards are evolving, so the information should not be taken as a definitive statement of current capabilities. From Tables 10.1 and 10.2 it can be seen that SCAI is currently less feature-rich than CSTA. However, unlike CSTA, SCAI does mandate a call model, which makes its behaviour more predictable as applications are moved between different vendors' switches. SCAI was designed with an emphasis on the provision of network services, and is considered by some to be the more secure protocol. However, a full implementation of the SCAI standard has never been brought to market, although a number of partial implementations do exist.

Most CTI applications are developed on a telephony API supplied by the computer vendor. These are almost invariably proprietary, although Novell's TSAPI is gaining acceptance as a *de facto* standard.

Such APIs have the effect of shielding the applications developer from the detail of the physical CTI interface, and, in doing so, make the application portable across the range of switches supported by the chosen computer vendor. Unfortunately, this switch independence is gained at the expense of lock-in to the computer manufacturer, and, to an extent, this negates the advantages of standardization of the CTI interface.

Some software suppliers have adopted the approach of developing an API which exactly mirrors the CSTA standard. Notable among these is Novell, whose product (TSAPI) also follows the trend in 3rd-Party CTI systems towards client/server architectures [2].

Figure 10.6 illustrates a typical client/server CTI installation, and it is seen that a telephony server (possibly duplicated for increased availability) distributes the switch's capabilities to a number of distributed client applications. These might be providing system-wide facilities (such as call-logging, call-distribution, or routeing services), or could be running call-handling applications on a user's workstation. In the latter case, these applications will look very much like those which would be provided through 1st-Party CTI. However, they will benefit from reduced costs in a large installation, since the cost of the server is likely to be outweighed by the saving of a telephony card per PC.

Currently, most 3rd-Party CTI systems are used in call centre applications, usually for inbound or outbound telemarketing operations [3]. An examination of the commands and events supported by CSTA and SCAI (such as Agent Logged-In/Logged-Out/Ready/Not-Ready) clearly indicates that the standards have been

Table 10.1   Showing a mapping, at the functional level, of the various host to switch telephony services defined within SCAI, CSTA and CS-1. (Note that services are likely to be similar, rather than equivalent.)

SCAI services [host→ switch]	CSTA services [host→ switch]	CS-1 operations [SCF→ SSF, SDF and SRF]
Initiate_Monitor (parameter defines event 'mask') Change_Monitor_Filter	Monitor_Start_Service (parameter defines event 'mask') Change_Monitor_Filter	Request report BCSM event Request every status change report Request 1st status match report Request notification charging event
Query_Monitor		Request current status report
	Query_Device	
Cancel_Monitor	Monitor_Stop_Service	Cancel status report request Cancel Activate service filtering
Make_Call	Make_Call	Initiate call attempt Connect Establish temporary connection
Route_Selected		Select route Select facility Connect to resource Collect information Analyse information Apply charging
Answer_Call Clear_Call	Answer_Call Clear_Call	Release call Disconnect forward connection Continue Call information request Furnish charging information Send charging information
Predictive_Make_Call Conference_Existing_Calls Drop_Conference_Party Single_Step_Transfer	Make_Predictive_Call Conference_Call Clear_Connection Consultation_Call followed by Transfer_Call	
Consultation_Transfer Hold_Call Retrieve_Call Query_Feature Set_Feature Routeing_Toggle	Transfer_Call Hold_Call Retrieve_Call Set_Feature	Hold call in network (similar)
	System_Status Event_Report Call_Completion_Service Reconnect_Call Divert_Call Snapshot_Call Snapshot_Device Alternate_Call Escape_Service	Call gap Reset timer
	System_Status	Activity test Play announcement (SRF) Prompt and collect user info (SRF) Query (SDF) Update data (SDF)

Table 10.2   Showing a mapping, at the functional level, of the various switch to host telephony services, events and triggers defined within SCAI, CSTA and CS-1. (Note that services and events are likely to be similar, rather than equivalent. In particular, the call control function will suspend processing when it reports a CS-1 trigger, whereas a CTI call control will continue after reporting an event.)

SCAI services and events [switch→ host]	CSTA services and events [switch→ host]	CS-1 services and triggers [SSF, SDF and SRF→ SCF]
Monitor_Report information Route_Request	Route_Request Route_Used Route_Select Route_End Re_Route	Analysed information
		Specializedd resource report(SRF) SDF response (SDF) Event report BCSM (parameter indicates event) Initial DP
Service_Initiated	Service_Initiated	
		Originating attempt authorized
Call_Originated	Originated	
		Collected information
Call_Delivered Call_Arrived	Delivered	
Call_Received Call_Established	Delivered Established	O Answer, T Answer*
		O Midcall, T Midcall*
Call_Cleared	Call_Cleared	O Disconnect, T Disconnect* Event notification charging
Call_Failed (parameter indicates cause)	Failed (parameter indicates cause)	O No Answer, T No Answer* O called party busy, T called party busy* Route select failure Call information report
Agent_Logged_On Agent_Logged_Off Agent_Ready Agent_Not_Ready Agent_Working_Not_Ready Agent_Working_Ready	Logged_On Logged_Off Ready Not_Ready Work_Not_Ready Work_Ready	
		Service filter response
Calls_Conferenced Conference_Party_Dropped Call_Transferred Call_Held Call_Retrieved Call_Diverted Network_Reached	Conferenced Connection_Cleared Transferred Held Retrieved Diverted Network_Reached	
		Status report
	Queued Back_In_Service Out_Of_Service Call_Information Do_Not_Disturb Forwarding Message_Waiting	
		* (The O and T prefixes indicate whether the event is reported at the originating or terminating end of a call)

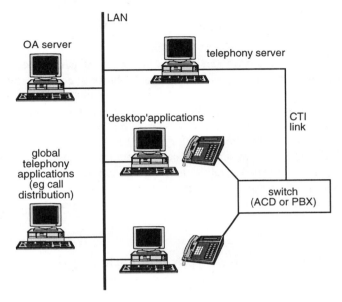

Fig. 10.6   Client/server CTI architecture.

heavily influenced by this very important and fast-growing market. The benefits brought to a call centre through the inclusion of CTI include:

- the ability to use multi-skilled agents — by using CTI to present the dialled number to the telemarketing/scripting application, the agent will be presented with the appropriate screens and scripts for handling the requested service (e.g. sales, enquiries, service);

- the presentation of calling line identity (CLI) to the application allows the automatic selection of the (likely) customer's records;

- the compilation of sophisticated call-handling statistics (the life-blood of the call centre manager);

- the ability to be able to simultaneously transfer a call and the associated data (customer records, order forms, etc).

These in turn lead to cost savings through increased efficiency, improved customer/caller satisfaction, and increased job satisfaction for the agents.

Although call centres have been the primary focus for the development of 3rd-party CTI, the capabilities of these protocols are generic enough to be able to build a wide variety of services, and there is a growing interest in services such as 'hot desking' and 'personal numbering'.

10.4.3   Observations

With the massive variety of facilities available on modern private switching systems largely exceeding an individual user's ability to comprehend, let alone use them, there has been a continued drive up the value chain into enabling a business's 'back-end' data to be tightly coupled into 'front-end' office applications. In many ways this can be seen to be indicative of where IN-based public networks will end up once the major services have been defined and deployed. However, this may be a somewhat restricted view of the future since there is a largely untapped potential in bringing the two technologies of IN and CTI together [4, 5] and exploiting the relative strengths of each, particularly as each environment makes use of increasingly similar distributed computing platforms.

10.5   RELATIONSHIPS BETWEEN PUBLIC AND PRIVATE NETWORKS

As already indicated, CTI protocols are generic enough to be able to build a private equivalent of many IN call-handling services. The removal of restrictions on network-to-network calls also mean that such services are not necessarily restricted to the private network domain, and there are examples of 3rd-party service providers using CPE to deliver facilities which would normally be considered to be in the domain of the telco. Clearly, these services will suffer an economic penalty when they require calls to be 'tromboned' out of and back into the network, but this is not always the case. As an example, consider a 'personal numbering' service implemented on a corporate network. Given that the personal number would ideally apply to all calls, including those which originated within the private network, then an analysis of the possible call sources and destinations will indicate that for most users there will be little difference in the call charge costs for the private and public implementations. However, the private implementation has the potential benefit of being easier to integrate with other office automation systems, such as diaries and schedulers.

A comparison of the CTI and IN capabilities in Tables 10.1 and 10.2 indicates the similarities in the functionality offered in both environments. However, CTI in its various proprietary forms is a more mature technology, particularly in the USA, where it is the fundamental building block of many very large call-handling operations. Also it will not be easy for IN to close the gap on CTI, since the CPE environment is one in which responsiveness and innovation are easy to achieve. This environment is largely unregulated, is not constrained by licence restrictions, is fiercely competitive, does not incur the availability, process, billing and security overheads which apply to a network operator, uses cheap technology, and is a very easy area within which to develop.

However, the CPE and network environments should not be regarded as competing with each other, and a closer examination of the CTI and IN protocols presented in Tables 10.1 and 10.2 shows that they have been specified with entirely different objectives. The role of the network is to deliver calls, and the role of CPE is to handle them efficiently and in a user-friendly way at the network termination point. Thus a protocol like INAP does not assume capabilities such as transfer, call distribution, predictive dialling, etc, within the switching fabric, since these are in themselves IN-based services. On the other hand, while CTI provides a pragmatic interface to a switch or network which is already highly featured, it lacks the range of security, charging and other features required for control of a public network. By contrast, the IN has a range of functionality essential to the service provider (operations and maintenance, mobility, charging, etc) which is either poorly specified, or non-existent in CTI. CTI and IN technologies may therefore be considered to present orthogonal control views on to the public network with IN presenting a 'vertical' view for call delivery services and CTI presenting a 'horizontal' view for post-call delivery call-handling applications.

If either the network or CPE tries to compete in the domain of the other, then it is likely to lose. Abbreviated dialling is a simple but good example of a service which is far better provided from CPE than from the network. Equally, though, incoming call distribution between several locations is better performed from within the public network as this avoids CPE 'tromboning' calls back into the public network, and allows truly global distribution.

To an extent, the threats to the network from intelligent CPE are balanced by such opportunities. To take the earlier example, a corporate network provided via centrex services will be able to provide a truly universal personal numbering capability to its users without ever incurring double-switching charges. It is also seen that call distribution is another significant opportunity. Many large telemarketing operations are distributed across a number of sites, and there is an increasing interest in the use of teleworking in this environment. The ubiquity of the network provides the opportunity for truly global call distribution, with the efficiency and management benefits of being able to control this through a single (logical) queue.

The greatest benefits are achieved when the true roles of network and CPE are respected, and each is used as a component in the delivery of the total solution to the customer.

However, to achieve these benefits, there is a need to look beyond the purely telephony-signalling capabilities of the access network.

10.5.1   Relationships between public and private IN

It is apparent from a consideration of the architectures of both public and private IN systems that there are considerable similarities between the major elements of each architecture. At a simplistic level, each architecture comprises a switching system which is controlled by an external computing platform. To this may be added a number of supporting resources such as communications peripherals (voice response units, facsimile modems, etc) and database systems. These features are shown in Fig. 10.7.

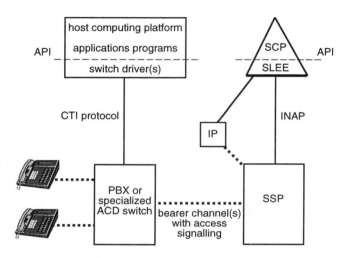

Fig. 10.7   Comparison of public and private IN architectures.

As there are such striking similarities between these two architectures, there are a number of potential opportunities for providing interconnection. The usefulness or otherwise of these interconnection opportunities is best illustrated through some examples.

10.5.1.1   Simple interconnection

In this case, which is typical of current call centre systems, only bearer-level relationships exist between the public and private IN platforms. Calls are processed in the public network before being delivered to the network termination point. Where additional call-related information is made available through the bearer signalling system, e.g. calling line identity, this information may be subsequently used by private IN systems to perform further independent processing on the call after it is delivered to the network termination point, but before the call is answered.

10.5.1.2   SCP access to private database

When executing service logic, a public network SCP platform will typically interact with a variety of data which is usually held in an on-line accessible database. A typical call flow through the platform will require one or more accesses to this data to route the call; for example a simple number translation application may 'dip' the database to obtain the translation of the dialled number. It is possible to extend this interaction beyond the public network platform's database to an interaction between the SCP and a database held within a private network. In this case, it is apparent that the access time to the database is a critical factor in the overall performance of the platform. Other factors which will be complicated by extending the interface between the database and the SCP to a database in the private domain include:

- security;
- reliability;
- availability.

These factors combine to make this particular interface potentially difficult to realize except where the service being realized can tolerate the delays and where the SCP technology can support the numbers of customers required. This last point is important in this context because a 'thread' of control may have to be maintained for a longer time period within the SCP for an external database access to the private database than would occur for a 'local' database access which implies the need for greater computational resources.

10.5.1.3   Host control of public switch

An alternative approach to providing private domain control over call delivery from the public network is to allow a private network host to control the calls delivered at a given network termination point. At the private network end, the structure becomes identical to a 3rd-party CTI structure with the PBX replaced by the local public switch and so this would be relatively simple to implement. Conversely at the public network end there is the need to ensure that the control partition is adequately policed to avoid the control interface acting upon calls which it has no right to manipulate (it is interesting to compare this with the early days of telephony where suspected partiality resulted in Strowger switching systems). The additional control flows between the local switch and the private host platform are also likely to generate a significant additional computational load for the switch processor (current services such as ISDN D-channel packet data and IN control signalling are notoriously computationally intensive for exchange processors) — particularly where several customers on one exchange require this type of service.

10.5.2   Similarities between CTI and IN architectures

The ECMA CSTA specification draws upon the open distributed processing work being conducted by ISO in identifying relevant functional domains [6]. With the view that the switch is the controlled entity and the computing platform is the controlling entity which hosts the various controlling applications, it is apparent that the computing platform must take responsibility for resolving the control flows to the various applications. For example, consider the case where a system uses two Freefone numbers — one for general customer services and sales and the other for servicing and maintenance. Incoming calls will cause 'route requests' to originate from the switch in response to received calls and it will be the responsibility of the computing platform to resolve which 'service logic' instance should be executed for each call. In this example the lower layers of the computing platform could route to either the service logic for customer services or the service logic for servicing and maintenance, dependent upon the called line identity.

In the case of public IN platforms, elements of the software functions which provide this routeing function are embodied in the service switching and service control functional entities identified in ITU recommendation Q.1214 [7]. By contrast CTI systems tend not to identify this functionality in such a generic manner as it is generally encapsulated within the proprietary applications environments which vendors provide on top of the basic CTI interfaces as indicated in Fig. 10.5.

However, there is a high degree of similarity between the control platforms for IN and CTI systems in that they must provide the distribution of control flows between the interface to the switch and the call-handling application. This call-handling application could also be on a separate processor to that supporting the switch interface and, in a CTI environment, this is almost certain to be the case. This could present difficulties for applications developers as they have to know the specific structure of the control platform in order to develop the call-time applications. This is clearly undesirable and fortunately this is a sufficiently generic distributed-systems problem to ensure that solutions either are available or will become available.

10.6   CONCLUSIONS

Although public and private telecommunications networks have common roots in manual switching, they have evolved into IN-type architectures for very different reasons. In the case of public networks this has been seen to be due to the need to provide a more flexible route to offering an increasing range of call delivery and handling services. Public IN developments have accordingly been directed at satisfying the need to support call manipulation at the pre-call delivery stage, with a relatively find degree of granularity. Conversely private systems have been driven by the user requirements of greater integration with both new and existing office/business automation and business information applications. CTI developments have therefore been driven by a slightly different set of requirements and in some cases the associated control protocols have been produced at a more coarse degree of granularity than those employed for public IN. There are also some special concepts embodied within CTI protocols which have no true parallel within public IN. These include the concept of supporting 'agents' in the context of telemarketing systems. Applications programming interfaces within CTI systems are also progressing towards *de facto* standards such as Microsoft's TAPI, whereas the near equivalent call-time SLEE interfaces on public SCP platforms remain largely proprietary and closed. In contrast, public IN systems have a greater integral support for operations and management than that offered by CTI systems.

Set against this background of contrasting drivers for public and private IN systems, there are striking architectural similarities between the two. These have been seen to establish similar requirements for the controlling computing platforms [6, 7] in terms of the need to support multiple call time applications to the switch side while maintaining a single consistent view for application developers and system operators. Both of these effects have combined to drive the computing platform solutions towards the use of a distributed systems approach to avoid data duplication and application incompatibilities. As both public and private systems migrate to such distributed control system architectures, there will clearly be increased scope for interworking and at some point in the future it may be practicable for the public and private systems to share a single holistic platform.

REFERENCES

1. Tanenbaum A S: 'Structured computer organisation', 2nd Edition, Prentice-Hall International (1992).

2. Wilson N: 'PC players set the CTI standard', Telecommunications, International Edition (October 1994).

3. Bonner M: 'Call centres — doing better business by telephone', Journal of the Institution of British Telecommunications Engineers, 13, Part 2 (July 1994).

4. Swale R P: 'Virtual networks of the future — converging public and private IN', IEE Colloquium on Virtual Networking (October 1993).

5. Swale R P et al: 'Convergence of public and private IN', Proceedings of the 2nd International Conference on Intelligence in Networks, Bordeaux, France (March 1992).

6. Computer Supported Telecommunications Applications — ECMA 179 (1993).

7. ITU-T Recommendation Q.1214: 'Distributed functional plane for intelligent CS-1' (1993).

11

PERSONAL MOBILITY SERVICES AND PERSONAL INTELLIGENT COMMUNICATORS

D G Smith

11.1   INTRODUCTION

Network intelligence concepts used in telephone networks are leading to a wide range of automated services that in the past could not be implemented without human interaction with the user. Personal mobility services are a family of services that are becoming increasingly important today. They are designed to match the needs of individuals who are mobile. In this context, mobile means any access mode may be used as desired, not only cellular mobile. Wired access or wireless access may be used from any terminal, in any combination with the same familiar services instantly available, leading to communication at any time, in any place, using any mode.

These concepts are not new. They form the basis of a set of standardized services called universal personal telecommunications (UPT), a key driver for intelligent networks. In UPT work, the terms personal mobility and terminal mobility are often used. Personal mobility refers to the user's ability to access telecommunications services from any network and terminal on the basis of a unique personal identifier, and the capability of the network to deliver those services according to the user's profile. Personal mobility concepts can be applied to both wireless and wired terminals, and result in a 'personalized' telephone. Personal mobility services are a vehicle for convergence between wired and wireless access, as illustrated in Fig. 11.1. Terminal mobility describes the user's ability to access services while in motion, i.e. while using wireless access, such as cordless, cellular mobile or satellite.

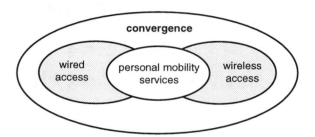

Fig. 11.1   Wired and wireless access service convergence.

Personal mobility implies a choice. Users can have telephony services tailored to match individual needs. Thus the user has many options from which to choose. Examples of this are variable incoming call routeing based on a personal number where the user defines the destination network number to be used when the personal number is dialled, or personal billing where an outgoing call can be made from any telephone and the charge billed to the user's personal account. The concepts can be further extended to allow service features to be enabled, disabled or modified as the user desires. For example, at certain times users may wish to be alerted when their bill has reached a chosen limit, while at other times this may be of no interest. The required service feature might be selected by the user and a limit entered, or deselected when not required. The information about a user's chosen combination of service features and options, and the necessary data, such as time or destination number, must be held in the network database. When changes are complete, users will expect them to have immediate effect.

These scenarios raise some interesting implementation problems, incuding those associated with user access to the database, which is a major focus of this chapter. Some important questions a user might ask about their service configuration data include the following:

- How do I know what options are available?

- How do I select the options I require?

- How do I know which options are currently selected?

- How can I cope with entering the many digits required?

The simplest services may offer only a small number of feature options, where one digit is sufficient to identify the option desired, but even the simplest services may require many data digits to be entered. Table 11.1 illustrates the point by listing the stages that might be used to enter the data digits required to configure a user's 'profile' for a schedule-based variable-routeing service. If short codes are used to select the first and second choice destinations, in the example shown, 54 digits would be required. If short codes were not used, 68 digits would be required. Neither is acceptable when using an ordinary telephone DTMF (dual tone multi-frequency) keypad.

Table 11.1 Example profile entry illustrating the large number of digits required.

User entry	Number of digits
Access (e.g. 0800 xxxxxx)	10
ID and PIN	20
Select option	1
Date	6
Start time	5
End time	5
First choice destination	3(10)
Second choice destination	3(10)
End	1
Total number of digits	54(68)

To select the desired option the user typically hears a voice announcement and responds by entering DTMF digits or by a spoken response. In the case of a spoken response, voice recognition in the network is required. In practice, it has been found that both suffer from being slow, leading to user frustration. However, it is possible to reduce the problem by keeping the DTMF detector on line so that voice announcements can be interrupted provided the user remembers which digit to press. Even so, it is still a lengthy process if there are more than a small number of options. An acceptable number might be two or three options.

As the number of service options increases, it becomes necessary to partition them into sub-menus, just like the menu-driven computer software packages widely used before graphical user interface (GUI) techniques were invented. A new problem is then found to occur when using an ordinary telephone. Voice announcements and selection digits have to be remembered, unlike when using a computer, where the menu can be viewed on the display screen. This becomes especially difficult when it becomes necessary to navigate a menu tree in order to select required options. Users often forget where they are in the tree structure, or they forget which digit they should enter next to select the option required. They are forced to traverse back up the tree again to find out what to do.

Having successfully selected the options required and entered the data digits, yet another problem occurs when users decide to make a new change. They cannot always remember which options were configured previously and which data digits were entered. It then becomes necessary to include voice announcements to read back the current profile, provide a printed copy by facsimile, or the user must write down each change as it is made.

These problems lead to user dissatisfaction. Using the service is not the pleasurable experience it should be. Instead of stimulating usage and creating revenue for the network operator the user interface problems become a barrier.

11.2   USERS IN CONTROL

Today, fixed network local exchanges support some services beyond the ability to make and receive basic telephone calls. They are specific to the access lines being used. Examples of these are:

- divert when busy;

- three-way calling;

- call charge advice.

They are invoked by entering a start digit (e.g. '*') followed by the code number for the service required and a closing digit (e.g. '#'). Analogue cellular networks offer similar services.

In the United States, customer-calling local area signalling system (CLASS) services for wired users have grown rapidly over the past decade. CLASS is defined by Bellcore as a method for transferring data between the local exchange and CLASS customer premises equipment (CPE). The services offered include calling number identification, calling name identification, automatic call-back and selective call forwarding. They use information delivered to the local exchange in the SS7 signalling channel and can be implemented using a traditional switch-based approach. They do not demand an intelligent network implementation.

Trials using screen phones have been carried out where options are selected by pushing buttons next to a list of choices shown on a display, instead of entering commands using an ordinary telephone without a display. Users find the visual interface much easier, instead of having to remember service codes and telephone numbers. Further evaluation of the results showed that an enhanced signalling protocol was required, leading Bellcore to develop a new protocol called the analogue display services interface (ADSI). Several companies are now testing ADSI telephones that allow customers to manipulate and control advanced services [1].

CLASS and ADSI have shown that more complex telephony services require a much greater level of user control than in the past, and that a simple telephone keypad without a display is not suitable. Personal mobility services incorporate a 'personalized' telephone concept, where a user perceives the same service independent of the terminal used. Screen phones can store user information such as telephone number lists, service access numbers and preferred control sequences.

If personal communications users were restricted to wired access in an intelligent network scenario, the screen phone approach offers a possible solution to the service control problem, but for the user who is not restricted to wired access it may be only a partial solution.

In the GSM system, subscriber identity module (SIM) cards are used to identify the user for billing purposes and to store personal information, user preferences and messages received via the short message service. A GSM telephone cannot be used unless a valid SIM card is inserted. This technique for achieving terminal independence could also be applied to wired telephones.

In future intelligent network implementations it can be expected that service control will be provided by communications between the service management system (SMS) in the network and service applications in the user's terminal. The service changes established via the SMS are downloaded to the service control point (SCP) for execution. Terminals are expected to employ visual representations and simple entry techniques that turn the service control problem into a pleasurable experience. Thus network intelligence can be combined with terminal intelligence to remove complexity and to make telephony service control easier.

Communications between the SMS node and the terminal application may typically take place via a modem located in the local exchange or in an intelligent peripheral (IP)/service node as illustrated in Figs. 11.2 and 11.3.

In Fig. 11.2 the network modem is co-located with the local exchange line card. A non-circuit-related signalling channel is used to carry information between the network modem and the SMS.

Figure 11.3 illustrates an alternative implementation where baseband modem signals are carried in a speech channel through the switch to the IP where the network modem is located. A non-circuit-related signalling channel is used to carry information between the IP and the SMS.

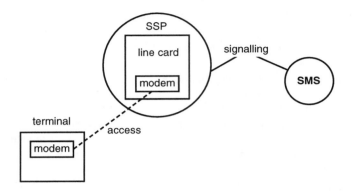

Fig. 11.2   Service control via line card.

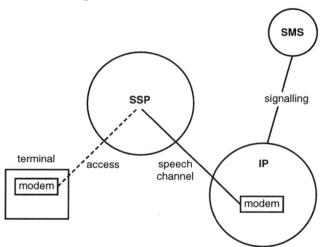

Fig. 11.3   Service control via intelligent peripheral (IP).

11.3   INTELLIGENT TERMINALS

The user control problems and service options lead to a vision for the future — a single communications device or terminal that can be used for both voice and data, incorporating seamless, fully integrated communications. The term 'personal intelligent communicator' or 'PIC' is used to describe this device. PIC characteristics include:

- seamless integrated communications access, wired or wireless;

- access to voice, electronic messaging and electronic information services;

- service feature control;

- a hand-held, pocket-sized device;

- highly integrated applications;

- a large display screen;

- a simple input device;

- long battery life.

In the PIC concept, intelligence in the terminal application software and intelligence in the network service software combine making the complexities of the technology invisible. The services become consequently easy and enjoyable to use. The concept is illustrated in Fig. 11.4.

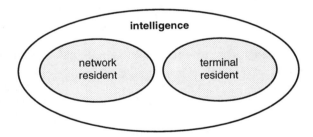

Fig 11.4   Combining network and terminal intelligence to remove service complexity.

In 1993 computer manufacturers launched a new kind of computer called a personal digital assistant (PDA) which goes part of the way towards the PIC vision. These are essentially electronic personal organizer products combining computing with data communications, but do not include voice telephony. PDA features include:

- integrated personal organizer type applications;

- communications by facsimile and e-mail;

- pen-on-screen input in place of a keyboard;

- handwriting recognition;

- intuitive user interface;

- large LCD display;

- infra-red port;

- pocket size;

- battery operated;

- Personal Computer Memory Card International Association (PCMCIA) card slot .

More recently, software agent concepts are leading to new proposals towards integrating terminal-resident intelligence with network-resident intelligence, giving a closer approach to the PIC vision.

11.4   PERSONAL DIGITAL ASSISTANTS (PDAS)

PDA applications may include communications, calendar, address book, note book, calculator, world clock, file manager, games, spell check and language translator. Applications have a graphical user interface designed to be intuitive and easy to use. They are described as integrated because applications can readily exchange information. Keywords can be used in one application so that information entered is automatically registered and available to other applications. For example, an appointment entered into the notepad using the keyword 'meeting' is automatically entered into the day planner at the time and on the day required. Similarly, if a person's name is entered into the day planner and their details are already stored in the address book, a simple gesture with the pen reveals the information.

Communications features can be used to create and send facsimile messages or to send and receive e-mail messages. Facsimile messages can be sent to any compatible facsimile machine, but to use e-mail an account with an e-mail service is required. The communications software allows simplified transfer of telephone numbers from other applications such as an address book, but the procedure for generating an e-mail or fascsimile and sending it often requires several steps that can be laborious. This is partly because the software is designed to communicate with what are essentially manual systems.

The PDA concept focuses on computing with communications, whereas the PIC concept focuses on communications with computing. At first sight this may appear to be only a subtle difference, but it is very important to recognize the shift of emphasis from computing to communications. All of the PDA features are equally desirable in the PIC, but with the addition of fully integrated communications hardware and fully integrated communications software applications in one case. Not only are the software applications integrated within the PIC, they are also tightly coupled with the communications access network and service features provided by network intelligence.

11.5   NETWORK SERVICES

The combinations of network-resident intelligence and terminal-resident intelligence leads to easy control of personal mobility voice services, but voice is only one of the communication modes of interest. PDAs and PICs are ideal for entering and viewing information and for creating and reading textual messages. With seamless integrated communications, mobile users can benefit from combining service control with personal mobility services and access to information, messaging and multimedia services.

Public information services today range from recorded voice announcements to on-line computerized services. Voice announcements are easy to use, but time consuming, expensive and offer minimal user control. Many telcos have tested interactive voice services using voice recognition to enhance user control. Voice is a slow medium and should be used appropriately. Computerized services are difficult to access for those who are not computer literate, and are costly to use. There are many information services that would be desirable, if only they could be accessed and controlled easily, and the terminals and services were priced for the consumer market. Examples include electronic books, newspapers, travel information and interactive books and games. In Europe mostly business and special interest groups use computerized services at present. There is increasing take-up of Internet services in the UK using personal computers, mostly desktop machines or notebooks. Users access the Internet via the fixed network (PSTN or ISDN), or, to a lesser extent, via digital mobile (GSM).

Like information services, many consumers do not use public e-mail services, probably for similar reasons of cost, accessibility and complexity.

Sending and receiving facsimiles electronically is largely an adaptation of a manual process. The widespread use seen today results from a high penetration of paper facsimile machines in businesses and a growing number of domestic machines and computers with facsimile capability. The limitations of point-to-point facsimile are well known. Inclusion of a facsimile capability in PDAs is perhaps a reflection of the delay that has come about in provision of low-cost, but potentially superior, electronic mail services.

Paging has been very successful for many years as a very low-cost short-message service. Separate pagers have often been used to complement voice communications, particularly for mobile users. However, the latest digital cellular technologies, such as GSM, have superior short-message service capabilities that remove the need for a separate pager.

The rate that information can be delivered to and from a PDA or PIC is dependent upon terminal capabilities, network service capabilities and access channel bandwidth. Both analogue systems and end-to-end digital systems have bandwidth limitations. Today, there is sufficient bandwidth for some services, such as travel information, but greater bandwidth and more local storage in the terminal will be required before more highly featured services, such as electronic book, newspaper and interactive services, can be implemented most effectively. The higher bandwidths required for multimedia services are expected to be realized when third generation mobile technologies are implemented, and as variable bandwidth transmission systems, such as, ATM become established.

11.6   INTELLIGENT NETWORKS

Intelligent networks can deliver complex personal mobility services with many features. Each feature may have a number of user-controllable options. Each option may require a large number of data digits to be entered.

Some control or data entry is possible using DTMF or spoken responses to voice prompts, but trials have shown that this technique is suitable for simple services only. Wired display phones using modem signalling can be used to improve the user interface, but they are not mobile.

Comparison of these solutions with a PDA or PIC clearly shows that PIC offers the optimum solution for mobile users because the key focus is communications, including voice. In view of the PDA features currently available today, there is an obvious growing disparity between the capabilities of intelligent terminals and the services offered by intelligent networks. Not only can the terminals be used for personal mobility service control, they are ideal for access to, and control of, messaging and information services. Intelligent networks have yet to exploit these capabilities. Terminals need not be restricted to one mode of access such as wired or cellular mobile. Users can choose whichever is most convenient or economic.

Intelligent network architectures of today have the potential to do much more than provide voice telephony services. They can enable seamless convergence of communications as perceived by users by integrating service control, personal mobility services, information services and messaging services, as illustrated in Fig. 11.5. The needs of single-mode access (wired or wireless) users and multimode access (wired and wireless) users must be fully addressed. An appropriate distribution of terminal-resident intelligence and network-resident intelligence is required. Conversion of existing networks to IN architectures, installation of new networks, or installation of overlays, is costly. It is important that the revenue-earning potential is fully exploited.

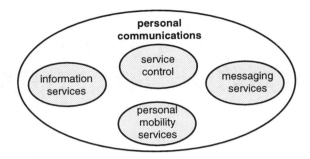

Fig. 11.5   The personal intelligent communications vision.

11.7   CONCLUSIONS

Currently, network intelligence is being used to implement increasingly more complex voice services in wired and cellular networks. Services are rapidly becoming too complex for users. New terminals, such as PDAs and PICs, incorporating intelligence built into soft-ware applications, can communicate with network-resident intelligence automatically, thus simplifying the user interface to the service. This combination of intelligence can greatly reduce the complexity of voice services as perceived by users, particularly when controlling or managing their personal communications. At the same time electronic messaging and information services can be readily accessed. Personal mobility services can form a bridge between different modes of access as illustrated by convergence of wired and wireless access services.

REFERENCE

1. Caruso R E and Gary J: 'Handler', Bellcore Exchange (December 1993).

12

ISDN AND IN WORKING TOGETHER

R G Buck

12.1   INTRODUCTION

The integrated services digital network (ISDN) and the intelligent network (IN) have grown from very different beginnings. Both ISDN and IN define a whole network solution but both have been designed from opposite standpoints. The ISDN standards were developed mostly from the view of the network-to-user interface, with heavily standardized services and protocols and little regard for the functions of the network elements. The IN was developed later, from the perspective of a deregulating telecommunications market-place, with companies needing to differentiate their network by providing services different from the competition. As a result, the IN standards discuss standardized service building blocks that can be put together in a wide variety of ways to allow this differentiation of services. To allow this type of service flexibility it was necessary to functionally decompose the network architecture and define the relationships and functionality of the elements within the IN architecture. As a result the IN and ISDN service offerings have grown apart, rather than together. This not only gives service problems for the customers but also represents an opportunity that has so far been missed. The ISDN has a very powerful digital bearer and signalling interface to the customer's equipment and a good network signalling system. The IN has its main advantages in rapid provision of complex services, and also has a network signalling system to match. Could the synergy of these two powerful systems be harnessed to provide a better service? This chapter will provide some views on possible ways forward to exploit this opportunity and will also discuss some of the current short-term opportunities.

12.1.1   Signalling

The signalling systems used in IN-based services and ISDN signalling are very different. IN-based services are designed around the use of a plain old telephony service (POTS) telephone connection, i.e. the telephone's capabilities are simply to signal to the network using in-band tones and to be able to receive in-band tones and announcements. An example of a service operation is the use of a chargecard. Firstly the user dials the number of the chargecard service; this triggers the local exchange to refer the call to the IN. At this point the IN will provide announcements to the customer instructing them to type in an account code and PIN. The customer depresses the numbers on the keypad and they are transmitted to the IN in-band. The IN receives the digits and uses them to access the customer database. More announcements and digit collection follow as the customer places the call.

Signalling in the ISDN is based upon different principles; the speech channel and the signalling information are separated into bearer (B-channel) and data (D-channel). This enables the signalling to remain separate from the speech, so that the operation of a service does not interrupt any conversation in progress. An example of this is the provision of call charge advice during a call. In an ISDN call the charge advice is requested and the information is returned in the D-channel signalling and displayed on a screen on the phone. This process leaves the conversation in progress on the B-channel totally unaffected, so if a facsimile transmission were in progress it would continue error free. The ISDN signalling does not rely on tones, rather it uses a digital communications protocol [1]. This protocol allows for communications between the local exchange and the telephone for the provision of services. There is also provision for communications between the telephone and other service provision machines, using the service access point identifier (SAPI). This is currently not used in most networks, although the use of this technique is discussed later.

Some, but not all, ISDN telephones have the capability to send in-band tones (MF4) in the B-channel, once the call is connected. This is useful in current systems to allow ISDN telephones to control IN-based services. This is necessary because local exchanges do not pass on digits from the ISDN protocols to the IN.

12.2   THE CURRENT POSITION

Currently there are many detailed challenges with the interworking of the ISDN and IN; this is largely due to the different design principles behind the two sets of standards. A few examples and a view on how the industry might move forward are given in this section.

12.2.1   Current problems

A simple example that appeared difficult to fix was that some IN-based services assume a speech path is available to the customer throughout the call set-up, for announcements and digit collection. However, the ISDN does not connect the bearer channel until the called party has answered the call. This prevents the IN from collecting digits, during the call set-up for ISDN customers. To cure this problem the service control points (SCPs) have been configured to connect the call to itself for ISDN calls. This allows the provision of tones, announcements and digit collection during the operation of the IN-based services. This fix works for voice calls, but has drawbacks for other categories of call. This is due to the SCP connecting the call, thus preventing the process of negotiation for type of call, e.g. swapping to a data connection.

Another problem with using services provided over an IN, via ISDN access, is simply how customers can provide information to the IN. POTS customers use tones (MF4) which are automatically generated when the phone buttons are pressed. In the ISDN signalling system the equivalent of an MF4 digit is the information message (INFO), which can be generated at any point during the call and could be used to operate IN-based services. ISDN phones could also produce MF4 tones in-band to operate IN-based services. As a result of there being two possible methods of signalling, manufacturers' implementations are very patchy, with some terminals producing MF4, some INFO messages and some both. Current IN systems are only designed to accept MF4 digits in-band, so a large number of terminals cannot access the services at all. Another related problem is that many of the terminals that do generate MF4 digits will only do so once the call has been connected, rather than during call set-up, when the IN may need them. Even with willing manufacturers, the terminal designs have to be updated and then the problem will only be corrected over a long period of churn in the terminal market.

Another difficult to-solve-problem is service interworking between IN-based and ISDN switch-provided services. There are cases of ISDN supplementary services with similar functionality to IN-based services. The problem here is that the two services may need to be operated in very different ways. This is confusing for the customer who may for example use the IN-based service at home and the ISDN-based one at work. Also, if the services are invoked together, there may be undesirable interactions between them. This is only one example — there are many other ways in which service interactions between IN-based and ISDN switch-provided versions can cause problems.

12.2.2   The way forward

Short-term solutions to IN and ISDN interworking problems are already manifesting themselves, but these solutions often have bad side effects. These short-term fixes will continue to proliferate until the convergence of IN and ISDN is handled at a higher functional level. All through the telecommunications industry, from standards, through telcos and equipment manufacturers, there needs to be a recognition that ISDN is just another way of providing telephony services to the customer. The ISDN service has options to use data calls and is, as a result, perceived by much of the telecommunications community to be a data service. However, many customers see it as just another form of voice telephony service and so expect to be able to use all their services on ISDN as well as POTS lines. This customer-driven need means that all new telecommunications services should be designed by considering the network holistically, i.e. services must be able to be accessed over all interfaces the customer could use.

12.3   THE FUTURE — OPPORTUNITIES

In the future, telcos will need to sell more advanced services to their customers. The principal reasons for this are the basic business drivers of increasing revenues and maintaining and growing their customer base. In practice, this means providing differentiation from the competition and being able to charge for these new services. If these new and more advanced services are to be used, they must be services that the customer wants to use because they add value to his day-to-day business. They must also be services that the customer finds easy to use, so that they will not be put off using them initially and will continue to use them regularly. The issue of ease of use is one that can be addressed by the technology, while the decision of what services the customer will want to use must rely on marketing information. This section discusses the opportunities for the future and how the power of ISDN signalling and the IN-based services might be harnessed together.

12.3.1   The requirements on signalling

It is important to have an overview of the requirements for signalling for services provided by the IN. The type and complexity of the telephone itself has a part to play in the signalling requirements. A feature-rich telephone may hold much of the service logic itself, whereas a simple terminal may need to be sent the full detail of what to display on its screen. Similarly the complexity and type of services being used will have an impact on the signalling requirements. Three examples are used here:

- chargecard;

- call waiting;

- advice of charge.

These examples will help to clarify the more generic needs imposed on the signalling.

In the example of a chargecard call the customer needs to be able to communicate to the network a series of numbers in a secure way, prior to dialling the destination number. The customer would like guidance in the use of the system and would ideally like to have an abbreviated method of entering the information. For example, by swiping a smart card and entering a PIN (personal identification number). The guidance as to the use of the service could be either text on a screen or a spoken instruction in-band or both. All the information for the provision of this service may be conveyed in-band without jeopardizing the integrity of the service. However, where a screen-phone is in use, it would be useful to be able to give on-screen and spoken guidance at the same time.

In the example of call waiting with calling line identity (CLI) (Caller Display™), the network needs to send to the customer the number of the party who is calling and give an indication to the customer to tell them of the call waiting. This should ideally be done without any interruption to the call in progress. The signalling requirement can be summarized in this case as an out-of-band short message service to the terminal during a call.

Imagine a future service where the intelligence in the network is offering an advice of charge for a videophone call. If the advice of charge information is inserted in-band, the video would be corrupted, so out-of-band signalling is again required during the call.

Another major factor affecting the requirement for the signalling to support the IN is the type of interactions that the customer will have with the network. Will the customer enter a series of numbers to answer questions posed about the operation of the service or will the customer expect to use a feature-key based system? That is, a system where the terminal being used represents the functions required by a special button to invoke that service. An example to illustrate this is the use of call waiting. To invoke the service a customer with an ordinary phone has to type in a series of numbers such as * 5 7 #. However, a customer with a network services phone would have a button labelled 'invoke call waiting' that would accomplish the whole task. In the first case (stimulus signalling), the signalling system is conveying one digit at a time, probably in response to a series of announcements from the network. In the latter case (functional signalling) the signalling system can convey one message to the network which has the whole meaning. It is clear that functional signalling has the capability to make network services far easier to use and some POTS phones are now being delivered with network services programmed into some of the memory buttons. This type of facility could readily be carried one stage further — the network sending the meaning of the buttons to the telephone as and when they were needed by the customer; therefore a button's function and legend will vary according to the stage of a call. This adds sophistication and ease of use to the network services but requires greater signalling capacity from the network to the telephone.

12.3.2   Signalling solutions

One way of providing the signalling link between the intelligence and the telephone is to use the functional protocol [2], which is partially standardized. This protocol provides for a non-call-related signalling path based upon the register and facility messages in the ISDN D-channel. The equivalent messages in the network signalling are also partially standardized, with the relevant elements reserved for future standardization. If this protocol were used it would provide a strong signalling pipe, with a standardized meaning, allowing telephone and software manufacturers to supply software to interwork with the provided intelligence services. There would, however, be a need to standardize an application layer protocol defining the way in which the telephones needed to respond to the intelligence commands. The downside of the functional protocol is that the pipe it provides is of limited bandwidth, being basically a (very) short message service. This bandwidth restriction limits the use of the protocol to the simple control of services and applications and would not, for example, allow the display of still pictures on the telephone.

Another approach to the provision of this type of service over ISDN is to open a separate virtual pipe within the D-channel. This is possible using the service access point identifier (SAPI) which is intended to allow the telephone to signal to several network service processors. Currently the international standards use this method of opening up lower layer pipes to connect to the local exchange call-control and to X.25 services. Another connection could be specified which provided a connection to the telco's intelligence layer. This would effectively open a clear pipe between the microprocessor in the telephone and the intelligence in the network, allowing an efficient computing protocol to connect the two together for the provision of intelligence services. The bandwidth of this pipe is restricted to the capacity of the network to handle the data, and limited to an absolute maximum rate of 16 kbit/s on an ISDN basic rate interface.

Most forms of network provision for use of additional SAPIs groom out the traffic close to the point the access line enters the local exchange. This prevents the capacity of the links being restricted by the exchange processor capacity and moves the information on to data connections more suited to the transport of this type of message. This has implications on the network architectures for the provision of intelligence services on the ISDN and on the exchange provisioning itself, which are discussed below.

12.3.3   Architecture

The IN standards and architectures have been explained in Chapter 1. This section briefly describes the implications of implementing separate signalling for services in an IN. The architectural implications of using a separate SAPI to connect the ISDN terminal to the SCP's service logic are the requirements on the local exchanges and the type of network that would be created.

Using ISDN protocols to bypass the local exchange with service-control information is a powerful way of fulfilling one of the IN goals, i.e. to move the developments into the more highly reactive information technology (IT) environment and away from the traditionally slower switch suppliers. This would allow enhancements to exchanges to be limited to those necessary to support SSP functionality for POTS lines. This would also take away a major threat to the integrity of the switches, i.e. as IN-based services take off, the switch processors will gradually have more and more processing to do, until either the processors have to be upgraded or the exchanges replaced.

The other major implication of grooming out these messages at the front end of the exchange is that a new overlay data network is needed for the provision of IN services. In practice this would be done by expanding and reusing current data networks. However, there would be implications in terms of cost and operational maintenance. A significant advantage would be that the data network would be free of most of the slow-to-change telecommunications standards and it would take the whole service provision into the more reactive IT world. This would allow the upgrade of the signalling network to cope with new types of services which have not yet been thought of.

Having discussed the removal of the signalling bottlenecks, using ISDN protocols, the next section discusses how the final bottleneck — the user interface — might be overcome.

12.3.4   User interfaces

In order to make services over an IN easier to use, improvements to the user interface must be considered. Currently it supports only voice announcements to which responses are made by strings of numbers pressed on an MF4 keypad. Many have realized that there are benefits from adding a display to the telephone, and that, by making that display large, there can be a new range of service possibilities. Indeed, screen-phone market research and trials are commonplace now in the US and, on the whole, the trials are being well received [3]. However, the trials are mostly based upon the presentation of the information to the terminal in-band, which interrupts the speech path that may itself be in use. A simple example of this is the caller display service (CLASS™ services) when used during call waiting. When a call waiting indication arrives the speech path is disconnected for roughly a second while the display information and bleep are sent in-band to the receiving customer. A shorter disconnection for the bleep only is repeated every 20 seconds or so until the incoming call is either accepted or rejected. The result of the disconnection and bleeps is to make the existing conversation more difficult to continue. This is a very useful service and the interruptions are essential for its operation. However, by transporting the information to the customer without the interruptions and bleeps, the service could be enhanced greatly. The effects of the in-band bleeps and disconnection are worse if the call is conveying facsimile or data since it will cause errors. The solution is to have an out-of-band signalling system to convey this information. The ISDN access protocol has a signalling channel (the D-channel) permanently available, out-of-band, on which this information could be conveyed. No standards explicitly exist to encourage the use of D-channel signalling for the presentation of services over an IN. However, it is a very powerful method especially when combined with screen-phone terminals to display prompts to the customer. Currently IN-based services depend upon the use of voice announcements asking the user to press numbers on a telephone keypad which pass in-band tones. The sequences of announcements and responses can become difficult to use. A simple example is given in Fig. 12.1 of the retrieval of a voice message from a network-based call answering service.

☏	0800 xxxxxx
⇒	Welcome to the 'Mobility co' personal mobility service
⇒	Please enter your pin number followed by a hash
☏	xxxx#
⇒	You have one new voice message
⇒	Press one to make an outgoing call
⇒	Press two to hear your voice messages
⇒	Press three to configure your services
☏	2
⇒	You have one new voice message
⇒	<voice message played>

Fig. 12.1   An example of message retrieval.

This interface is usable, but it has to be regarded as less than ideal. There is a distinct limit to how many messages can be listened to before the customer starts to forget the earlier options. This is creating a bottleneck between the intelligence in the network and the customer, which, if removed, would allow far easier use of more complex services. A stronger link would allow more information on the use of the service to be passed to the customer and the customer to pass more information back to the service logic of the intelligence. This in turn would help to make the services far easier to use and would result in the development of far more powerful, but still easy to use, services. Figures 12.2 and 12.3 show how the same service could be presented by using a screen-phone interface. The

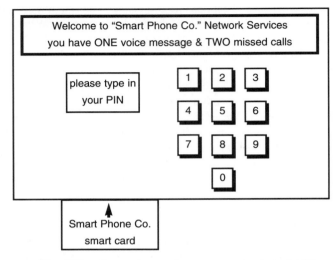

Fig. 12.2   Screen-phone — insert smart card and type in PIN.

screen-phone consists of a handset and a large display which is touch sensitive. This means the network can program buttons to be anywhere on the screen, with text or graphics in them, and the user simply touches the button on the screen. This gives an easy-to-use access medium for the integrated display of both information and IN-based services. Compare the example in Fig. 12.1 to the use of a screen-phone interface for this type of service, as shown in Figs. 12.2 and 12.3.

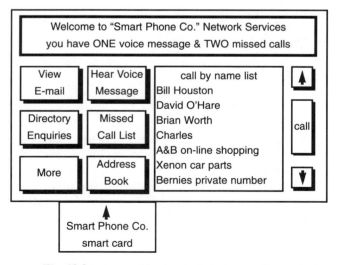

Fig. 12.3   Screen-phone — select whichever option required.

The diagrams of the screen phone in Figs. 12.2 and 12.3 are self-explanatory, showing how easy it could be to use an interface of this type. Simply inserting a smart card and typing in a PIN gives access from any suitable phone to the customer's own services. Then, pressing one button selects the service and allows the voice messages to be played. This type of screen interface could provide an integrated, easy-to-use access to all IN-based services and could also provide an excellent window into what are currently known as information services, such as local information pages with train and bus times, booking of cinema tickets, home shopping, home banking, etc. By using a good-sized screen on a phone, the interface to such services can be very powerful and very easy to use. Imagine seeing a plan of the cinema, on screen, before choosing which tickets to buy, or a full bank statement on the screen with the ability to look in more detail at particular transactions on-line. Many other information services are possible and would be available in a simple, easy-to-use format without having to own and understand a personal computer (PC). The result is that these services would be available to many more people than the small percentage of home computer users. This increased market penetration could enable this type of service to take off quickly. A relatively low-cost screen phone could get the information superhighway into homes and businesses very quickly.

12.3.5   The benefits of screen phones

To further illustrate the benefits of using a screen-based system with an advanced network intelligence, compare the following two short stories both about the same working day a few years in the future.

- Simon is a marketing executive for a big company. Each day he gets up at 6.30 a.m. so that he can have breakfast before he drives to the train station to catch the 7.36 a.m. train. He arrives in the office at about 8.25 a.m. Simon has recently bought a mobile phone so that he can keep in touch while he is on the train and has subscribed to some of the new services from 'A. Telecom' which allow him to divert his phone calls from his home phone to the mobile. His work PBX allows him to forward his calls from the office to his mobile as well, so Simon feels that he is now always in touch with his customers.

It's Tuesday morning, 6.20 a.m., Simon wakes abruptly to the sound of a bell ringing, he reaches for the alarm clock to turn it off, the bell continues. Startled and suddenly wide awake, Simon struggles with the alarm clock for a second or two before reaching across to pick up his mobile phone and answer the call. "Oh! Hi, Simon, it's Larry, I didn't expect you to be at work at this time, I was just going to leave a message on your machine." "Argh" replied Simon, "I've got my office phone diverted to my mobile number." "I suppose you thought that was clever," retorted Larry; "I was just ringing to let you know that Barbara has just gone into labour, I'm at the Hospital now, so I won't be in today — is that OK?" "Oh yes, fine, good luck! See you in a few days, I guess?" "Thanks, yes err.. bye then." Simon puts the phone down, and collapses back on to the bed. Three minutes later a bell rings again. Simon reaches out of bed and throws the mobile phone into the linen basket; the bell continues to ring, but now much louder; startled and suddenly awake Simon turns off the alarm clock, gets out of bed and retrieves the mobile phone. Later Simon picks up his mobile and as he leaves the house he types in the 18 digits from his crib sheet that diverts his home phone to his mobile. He drives to the station and gets on to his usual carriage of the train. First thing today Simon has an important meeting, so he wants to read up the paperwork on the train. The papers are still in his office because he forgot to bring them, so he reads the sports pages as usual. He is not particularly interested, but it passes the time. On arrival at his office Simon finds the lifts are out of order so he has to walk up to the 9th floor meeting room. He is a little late so is surprised to find nobody there. After a few minutes Simon becomes concerned, so he phones his office to find out what is going on. His mobile phone rings. "Typical," he mutters, slamming down the phone. He picks up his mobile but as he does so it stops ringing. The penny drops. Simon dials the front reception and is put through to his office, where Mary tells him that the meeting has been cancelled. "Didn't you get the message, I left it on the agenda on your desk?" Simon slams the phone down and walks down the seven flights of stairs to his office. When he gets there he finds Mary's message on top of the pile of papers he left on his desk the night before. "You look a bit out of breath," quips Barry; "Donald from A&B leisure came in a couple of minutes ago, wanted a quote, so I took care of it — lucky I was here otherwise we would have lost his business." It had always annoyed Simon that Barry always seemed to be in the know about everything and this just reinforced his annoyance. Barry is the same grade and works on the same type of accounts as Simon, but seemed somehow to be doing a bit better at it.

Barry and Simon's lives were actually quite similar, both living a similar distance from work and travelling in by train.

- Barry also got up at about 6.30 a.m. This morning when he got up he looked at his screen-phone in the bedroom and noticed that he had a message. He pressed the message button and saw that it was from Larry. It was timed at 6.18 a.m. which he thought was strange, but he noted that it had come in as a business call so he knew he could handle it later. Over breakfast Barry always checked his paper mail, and usually had a quick look at his electronic mail using the screen phone on the breakfast bar in his kitchen. He pressed the e-mail button and saw a note from Mary entitled 'meeting postponed'. He scanned through it on the screen and noted that the early morning meeting had been postponed until the afternoon. He tapped a couple of soft keys on the screen to store the message as a reminder to re-arrange his afternoon meeting. He looked at another electronic message from his credit card company, this time a bill. He read down the list, on screen, of things he had bought, before authorizing payment of the bill from his current account. He finished his breakfast and on his way out to the car picked up his personal digital assistant (PDA) from its charger and put it in his pocket. The charger notified the fact that he was now mobile with the network intelligence service from 'B. Telecom'. According to his personal profile, set up with 'B. Telecom' service, all calls were now automatically directed to a voicebank service for 10 minutes while he drove to the station and parked the car. Once on the train Barry felt happy to take calls again so the service directed all the calls to his PDA, which had a digital mobile phone built in to it. Barry sat down at his usual table and went through his usual routine of checking his daily appointments and messages that had been downloaded to his PDA. He checked the voice message from Larry and sent an e-mail reply from his PDA keyboard wishing him good luck with the birth. He then went on to re-arrange his afternoon meeting, using the PDA's capabilities to log on to the office network. While he was logged on he downloaded the summary of the papers for the morning meeting which Simon had forgotten to copy to him. Barry spent the rest of the journey reading the summaries and making a few notes for the meetings.

As Barry climbed the last few stairs towards the office he saw Simon running up past the office. He shouted out to tell him that the meeting had been postponed, but Simon was already out of earshot.

This story highlights a few examples of what a well-integrated personal mobility service on an intelligent network with a screen phone and PDA interface can do for customers. Some of the facilities might sound a little far fetched, but all could easily be possible with the enabling force of a screen-based interface to an intelligent network and sufficient signalling power to enable the transfer of the data. Indeed some features not mentioned in the story, such as home shopping, would be provided complete with good quality pictures of the items being bought.

12.4   A FIRST STEP INTO MULTIMEDIA AND INFORMATION

It has already been illustrated that screen-phones form a first step into multimedia and information services. In their most basic form they allow simplified explanations of the IN-based services and give text-based access to information services, such as bus and train timetables and simple home shopping. This in itself is a first step into the information arena, but, with the greater bandwidth available from ISDN signalling, such screen phones can offer graphical representations, pictures and even video calls. A typical application might be home shopping, where text and graphical menus allow quick and easy selection of products, which can then be viewed as still pictures or a short video clip. Another service might allow a look-up of train times and delays, on screen, while arranging a meeting over the voice channel. All of these types of services are possible and form a valuable first step into the information revolution. Screen-phone customers will be able to use all their advanced services without having to be tied to a PC or a VoD terminal. Many of these services can be provided via a PDA or PIC (see Chapter 11). Information services such as those described could be provided over many different types of equipment, with differing capabilities. In an office, services could be accessed from a PC, at home via a screen phone or a VoD terminal, and while on the move via a PDA. Each type of terminal will have its place in a portfolio of products to enable access to information services. Much of the technology is available now and could provide an early introduction of information services. This is possible using the narrowband network, without waiting for broadband technology to take off.

12.5 CONCLUSIONS

This chapter has shown that the combination of the power of the signalling and data capabilities of the ISDN network and the services provided over an IN are complementary. It should be seen as a powerful and lasting relationship, bringing easy-to-use intelligent network and information services to all customers. This combination of ISDN and IN technology would allow the integrated presentation of advanced intelligent network and information services without waiting for the take-off of broadband services. This is a technology that could be put into place in a short period and would be a first and relatively inexpensive step into information and multimedia services. This would, in itself, open up a change in culture and help to pave the way into full interactive broadband information and multimedia services.

REFERENCES

1. ITU-T Recommendation Q.931: 'Digital subscriber signalling system No 1', (November 1988).

2. ITU-T Recommendation Q.932: 'Generic procedures for the control of ISDN supplementary services', (November 1988).

3. Information Networks, 7, No 16 (August 1994).

13

TWENTY TWENTY VISION — SOFTWARE ARCHITECTURES FOR INTELLIGENCE INTO THE 21st CENTURY

P A Martin

13.1   INTRODUCTION

The communications networks of the future will support high-bandwidth, low-delay information flow. The likely application areas to take advantage of these gigabit per second networks are video-on-demand, animated shared simulations and globally distributed computing. As corporations grow to attain global reach, the scale of their computing and communications infrastructures will also grow. This chapter explores potential future software architectures that would support these new widely distributed applications. It then compares the needs of such architectures with the promise held in emerging distributed computing technologies.

The software architectures proposed are based on the assumption of computer-level integration of the communications industry and their customers. For the purposes of this chapter, customers as viewed by the communications industry come in two types. The first is the end user and is the consumer of services. The second provides a communications-based service to its customers, who are the previously described end users. This group is referred to as the service provider. This chapter assumes that service providers will have access to facilities that have previously been beyond their control. For instance, they will be able to establish connections between interested parties and to control the flow of information between them.

The chapter then goes on to explore what these groups have in common and seeks to model them as enterprises. The use of large shared-object systems to model an IN service provision enterprise is explored in detail. The examples given are set in traditional telecommunications intelligent network applications. The use of such object systems is as applicable to communications enterprises as it is to service providers or any other enterprise. Large shared-object systems are expected to form the fundamental building block of computer infrastructures of the future.

13.2   ASSUMED PHYSICAL ARCHITECTURE

It is not the purpose of this chapter to suggest divisions of responsibility between different enterprises contributing to the provisioning of future IN services. Rather the focus is on flexible software architectures that can be used in any division of responsibilities. Accordingly the following description of an assumed architecture is more by way of example. Figure 13.1 shows a likely overall physical architecture and each domain is described below.

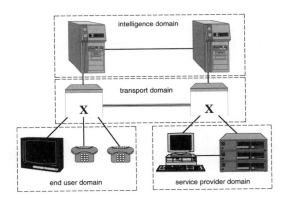

Fig. 13.1   Assumed physical architecture.

13.2.1   The role of the transport domain

In the centre of Fig. 13.1 is the transport domain which includes the physical network infrastructure, such as switches and fibre-optic links.

This domain performs the task of delivering information from one end-point to another end-point. The delivery can either be connection-oriented or connectionless. With connection-oriented delivery streams of information flow from one end-point to the other — so this is typically suited to playing a film once the equipment at either end is set up to either play or receive the flow. Connectionless delivery is used more in the messaging that is used to set up the connection. Accordingly the transport mechanism must be able to support both delivery methods.

For the purposes of the chapter, this domain includes the processing that controls which end-point is connected to which other end-point and with which bandwidth. This connection control activity is similar to call control in existing networks. However, the call concept is at a higher level and more telecommunications-specific. Software that uses connection control, such as the intelligence software, may specify end-points to connect between, and be informed of the progress of the connection.

13.2.2 The role of the service provider domain

The service providers are enterprises who advertise their services. They then have equipment ready awaiting connection requests. End users then dial in and connect to the service providers' equipment. A typical role would be providing video services. Accordingly they are shown in Fig. 13.1 with a computer and a video player. Having connected to a specific service provider, the end user would then interact with an application running on the service provider's computer in order to select a film. The appropriate bearer would then be set up across the transport domain and the film played to the end user's TV.

Traditional IN applications or services such as number translation and time of day routeing revolve around basic telephony linked to personalized customer data. Where do such applications fit into the software architectures of the future? Are they applications to be provided by the service providers at the network edge or are they so closely linked to the switching that they have to be near the core in the intelligence? Several aspects of this are discussed in Chapter 10.

13.2.3   The role of the end users

End users are similar to service providers in so much as they are both customers of the transport provider. They connect their equipment into the transport infrastructure and then use it to dial up services as provided by the service providers. Their role is therefore to consume services and transport bandwidth. Figure 13.1 shows end users equipped with appropriate apparatus such as television sets.

13.2.4   The role of the intelligence domain

Having described the role of the other three components of Fig. 13.1, the question arises: 'What remains for the intelligence domain to do?'

The times at which the intelligence domain becomes active seem to revolve around call set-up and clearing which may provide a rationale for deciding where specific services are best located. The traditional IN-based services, such as number translation and time-of-day routeing, which come into play before a call is established fit naturally into the intelligence domain. In addition to these well-understood applications other opportunities exist for the intelligence domain to make use of its position between end user and service provider.

The intelligence domain is aware of the other domains and can hold the necessary information to connect them together. The intelligence domain may provide advertising facilities. Service providers would advertise their services and end users may view and select from them. This is similar to the current 'Yellow Pages' service. Having located the appropriate service provider the intelligence domain would route further signalling through to them. The selection of services is likely to be automated to some degree in the future, making use of personalized software agents. An end user would load its agent with a query and it could inquire of various service providers about the options available. The results would then be delivered back to the end user for them to select an option.

Also, given numerous service providers wishing to bill their customers, an opportunity exists for intelligence to offer integrated billing services.

13.3   A SOFTWARE MODEL FOR INTELLIGENCE SERVICES

The software architectures that are best suited to controlling these systems are now assessed. The foregoing text described a number of groups which use the transport domain. These are end users, service providers and intelligence. What these groups have in common is that they can be viewed as separate enterprises with their own supplier-customer relationships. This chapter seeks to recognize what enterprises, in general, require of a computing infrastructure with special emphasis on the needs of IN. In order to understand the requirements that enterprises expect their suppliers to fulfil, it is useful to develop a model of those enterprises.

Enterprises can consist of single individuals going about their business or, at the other end of the scale, they may be multinational corporations. From the communications company point of view, the enterprises that use their facilities must be understood so that any barriers to their expanded use can be removed. Communications companies are themselves enterprises and can also benefit from such an analysis.

13.3.1   Why model the enterprise?

Everyone who is involved in an enterprise carries a model of that enterprise around with them in their head. By example, if the enterprise is a window cleaner's round then the only people involved are the cleaner and the clients. The cleaner can remember the clients, their requirements and how much they owe. If anyone wants to know anything about the enterprise they must ask the cleaner. Sadly, few enterprises nowadays are this simple. A successful window cleaner would have taken on employees, more cleaners, accountants and the like. The model becomes more complex. External agencies, such as government departments, have to be comprehended and integrated into the model. If the enterprise becomes large enough, it is beyond the capacity of any one individual and this is where the trouble starts.

Typically, each part of the enterprise will have been divided up into separate traditional departments. The accounts department handles the finances and keeps their records based on the fiscal components of the enterprise. The roundsmen keep their records based on the requirements of their round, etc. These records have a number of components in common but are based on the functions performed by the separate groups. They are maintained separately and hence evolve into working systems with defined, probably paper-based, interfaces to other departments.

Automation of the record-keeping process usually serves to further define the functional, departmental nature of the enterprise. Islands of automation arise and interface systems are produced to provide integration. The result is that each enterprise maintains multiple computer systems each with their own overlapping models of parts of the enterprise.

It has therefore become evident that another approach could be more profitable. If it is possible to model the enterprise as a whole, then considerable benefits can be attained. For instance, any user can explore the running and data of the enterprise without the restriction of department barriers.

The situation is no different in the world of IN and service provision. Successful services have been installed as essentially stand-alone systems, needing interfaces to other systems. Telecommunications companies have been striving to introduce generalized systems and the remainder of this chapter suggests how this could be approached by modelling the provision of IN-based services in objects.

13.3.2   Why use objects for IN?

A number of technologies are available to enterprises that wish to model their activities. The dominant software methodology for enterprise modelling is likely to be object orientation (OO) [1-3]. Objects as a software concept are now entering the mainstream for new application development because they offer significant promise to overcome the barriers that hold enterprises back. Increasingly enterprises are restricted by their inability to quickly and safely evolve their software systems to meet new needs and opportunities. So what benefits do objects hold for enterprises in general and IN-based service providers specifically?

The most important benefit of an OO system is the encapsulation of software components. This is a form of automatic modularization that means the software component publishes a fixed interface. Behind the interface the actual implementation can be changed. This has the effect of dividing up the software into handleable-size chunks so that defects in one chunk do not pollute other software components. As, in reality, the definition of the published interface is rarely unambiguous, misinterpretations of an object's duties often lead to inelegant systems. This is more an issue of software engineer discipline than a failing of OO methodologies.

Another significant benefit of using objects is that they lend themselves to the modelling of real-life systems, especially systems that support the 'is-a-type-of' relationship via inheritance. Theoretically the one-to-one relationship between the real world and its model in software objects would lead to easy reuse of software components. There is currently little evidence of extensive and successful software reuse, but this is again an issue of software engineer discipline.

These benefits apply to IN applications as they do to any other application domain. Objects are therefore a powerful methodology to adopt for future systems, but strict control needs to be used in their deployment if their benefits are to be experienced.

13.3.3   Why use distributed shared objects for IN?

Having accepted the benefits of modelling the enterprise and using objects, then why add the extra complication of using distributed objects? The most compelling reasons in IN applications are model size and availability. Once an enterprise model gets to a certain size in terms of the memory, disk and CPU usage requirements it can no longer be hosted on only one computer. For example, an enterprise that provides IN-based services may have many millions of customers, each requiring their own personalized data to be ready for use whenever they invoke a service.

Additionally, in a single host system, if that one computer is unavailable, then the IN-based service as a whole is also unavailable and hence losing money. Distributed objects bring the promise of truly huge, continuously and globally available applications. Unfortunately this is only a promise at the moment as genuinely useful large object systems are not currently available.

An obvious but awkward requirement of large object systems is that many applications will be simultaneously accessing and editing the objects. For instance a billing application may want to read customer object data while a customer-handling application is updating it. Conflicts are bound to arise and the system should resolve these in an equitable manner without allowing the model to get into an inconsistent state. This means that the model has to include some form of object-locking or transactional system.

In addition to the general requirements for large shared-object systems, their use in IN applications applies extra criteria. These include high availability, near real-time response and very wide area distribution. These themes are picked up in section 13.4.

13.3.4   An example object model for intelligent networks

For a typical model of IN services the transport domain would support objects such as switches and ports. The objects supported by the intelligence domain would include devices, such as telephones and televisions, customers, calls and services. A simplified version of the model is given in Fig. 13.2, which shows the model that could be used to control the transport and intelligence domains of the assumed physical architecture shown in Fig. 13.1. It should be noted that there is a one-to-one mapping between a real world entity and an object in the model.

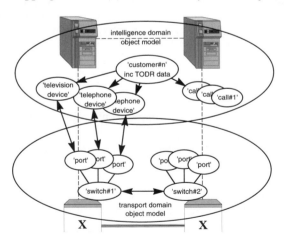

Fig. 13.2   An example object model for IN.

The transport domain would also support the switch and port objects. Each physical switch that the provider owns would be represented by a switch object in their object model. This means that the transport domain would need its own computing infrastructure. The switch and port objects could publish their interfaces to the other domains. Accordingly the intelligence domain could invoke operations on the switch object resulting in actions on the real switch and thus actions on the real world devices connected to the switches' ports. If the intelligence domain has some way of registering an interest in activity on ports, then the transport domain can invoke operations on intelligence domain objects. For instance, an off-hook event may be reported back to the intelligence domain's device object using this mechanism.

The device, customer, call and service objects would reside in the intelligence domain. The data associated with a customer's time-of-day routeing (TODR) is shown embedded in the customer object as it is uniquely associated with that specific customer. These objects could export their interfaces to the other domains. Service providers could use the call-object interface to request a path suitable for playing video across the transport network. Hence, the published object interfaces dictate the functionality that enterprises may request or supply to one another.

In this scenario, each domain or enterprise would have their own computing infrastructure connected to the transport infrastructure. On those computers they maintain their own enterprise models. At the edge of those models they can connect to other enterprises' models.

13.4   REQUIREMENTS OF A LARGE SHARED-OBJECT MODEL

The requirements for supporting large shared-object models are common to many application domains. This section points out the general requirements with emphasis on the needs of IN-based services.

There are three groups of people within each enterprise who would make demands on any system that supports object models. Any system that is to be widely adopted must satisfy their needs in a manner that is not too costly to the providers. These groups are the application users, application programmers and administrators. Their requirements are described below.

13.4.1   Requirements from the application user's point of view

13.4.1.1   Desire to create/destroy/read/edit objects

The application user will wish to interact with the object model as provided to them by the programmers. This would mean having the ability to create and destroy objects and to manipulate and browse those objects while they exist. To illustrate by example, application users could create customer objects, manipulate the attributes of that object and invoke its operations. When the customer takes his business elsewhere, the application user would delete the customer object from the system.

This activity is traditionally embedded in applications. If large-scale shared-object models are adopted, specific applications can be replaced with generalized graphical user interface (GUI) object editors or browsers. Attribute types, such as integers, floats and strings, can easily be represented in screen-based forms. These fundamental types will need to be augmented with new data types that are required to accurately model enterprises. These types could be accessed from the browsers and will include pictures, video clips and sound bites.

13.4.1.2   Continuous availability

Application users will want access to their applications at any time of the day, any day of the week and any week of the year. Traditional computer environments that provide continuous availability have relied on customized hardware that provides low performance at a high price. This price is handed on to the customers that demand the availability. From the application user's point of view, availability is calculated in terms of either unplanned time without access or absolute (planned plus unplanned) time without access. Obviously both types of downtime are undesired, but planned downtime is not as disruptive as unplanned downtime.

13.4.1.3   Ubiquitous availability

The application users will want to access their service from wherever they are. An enterprise that is multinational, even global, will expect global access from all its sites to all its sites. Moreover, the appliances that are used to interact with the enterprise model via applications will vary from site to site. Devices for access to applications will range from simple textual or voice to multimedia personalized devices. Increasingly these appliances will be evolutions of mobile communications and computing devices (see Chapter 11) [3].

13.4.1.4   Adequate performance

Application performance under 'hard' real time conditions has mandated criteria. Such conditions would typically be nuclear power station or airliner control, where failure to meet time limits would result in disasters. For most applications, such as IN-based services, the result of slow performance would be that a person or a system has to wait and the only consequence is that some time is lost. This does not mean to say that high performance is not desirable but any definition of concrete performance requirements must be treated with scepticism.

13.4.1.5   Security

The object model will contain objects and parts of objects that are not for general consumption. Enterprises will want to divulge parts of their object model to foreign users based on who that user is. For example, customers should not be able to edit the account component of their own customer object. However, they may be allowed to edit their own 'time-of-day routeing' data without service provider intervention. Accordingly the object model should be equipped with access control lists for each attribute of each object.

13.4.2   Requirements from the application programmer's point of view

The commonly used term 'application programmer' is probably too narrow in the sense meant here. The tasks undertaken by this group are defining and implementing the enterprise model in order to fulfil the needs of the enterprise. The skills required are commonly associated with analysts or designers. Any system that is to be widely adopted must please, even excite, this opinion-forming group. The system will fulfil the needs of this group in terms of the things they want to be able to do and, just as importantly, the things with which they actively do not want to be bothered.

13.4.2.1   Desire to create/destroy/read/edit classes

The enterprise model is likely to be implemented in class libraries. The definition of classes, what data they hold and what operations they perform, is a highly skilled task. Each class should model some concept in which the enterprise has an interest. For example, a communications company enterprise model is likely to include classes that define device and switch objects.

The programmers will want to be able to create new classes easily, to extend the object model. They will also want to be able to add new attributes and operations to existing classes. Similarly, non-useful class attributes and operations should be easily withdrawable. Although easily expressed, this is a very difficult need to fulfil. Once an object model is active, evolution of that model is very difficult in practical terms. The support for model evolution has to be built in from the earliest stage of system definition.

13.4.2.2   Location transparency

The programmers will desire location transparency which means they do not have to undertake any specific coding to ensure that their objects reside on the most appropriate hosts. Given that the object model will be made available via some computing fabric, then the programmer does not wish to give explicit instructions on how best to take advantage of that fabric. An object system could field all requests to create new objects and locate them on the most appropriate host.

The most appropriate host could be chosen on the basis of a number of criteria which are likely to be related to load balancing. A host would be selected if it has sufficient memory and disk space, if it has access to the object's run-time code and, finally, if it is faster than any other host that meets the previous criteria.

An object can be used to represent an individual and would contain many attributes. These attributes could include medical records, tax records, etc. All these attributes apply uniquely to the same person and therefore should be modelled as a single object, but it can be seen that it may be necessary to distribute them across multiple hosts. The tax authority will demand that the tax records are held on their secure hosts but may be accessible from the larger object model. Similarly, medical records, including complex data types such as X-ray images, video clips and speech annotated notes, may be held on separate hosts.

This requirement adds another level of complexity to the software if it is to be supported transparently. Alternatively, the application programmer can accept the necessity for attribute distribution and code the class libraries accordingly. This would mean defining 'medical records' as a separate object and having a reference to it from the person object.

Having accepted that an object can live anywhere, and also individual attributes of objects can be dispersed, there are conditions under which the choice of locations should be restricted by further qualifications. Switch objects may only be allowed to reside on computers that are physically linked to the switches they are controlling. Expressing the qualifications and restrictions on the locations of objects will present an awkward problem for system developers if they are attempting to achieve location transparency for the programmer.

13.4.2.3   Failure transparency

The programmer does not wish to pollute the class library source with exception-handling code for cases where a remote computer has failed to fulfil its duties. The programmer therefore desires to have failure transparency so that the system will catch and recover from failures without programmer intervention. This is obviously easier to express than fulfil. The number of potential failure modes for a single computer running a single process is large. Programmers normally have to add exception-handling code for situations such as memory exhaustion or accept the inevitable crashes. As IN-based services require high availability, the crashes must either be avoided or caught and fall-back mechanisms invoked.

When the scenario includes multiple processes on multiple hosts on multiple networks, the combinations of failure modes grow exponentially.

From the point of view of each object sitting within a process the situation is more simple. Each object, if it is a true object, has a message-based interface, i.e. all information and stimuli it receives are via messages. If it invokes a method in a remote object, it does this by passing it a message and then it will expect a response via a message.

If a computer crashes, the objects residing on that computer become permanently unavailable. It is difficult to recover from this situation, but not impossible. The system may be able to trap the crash event and replicate the objects elsewhere. From the programmer's point of view no extra code has been undertaken and failure transparency would have been achieved. Studies along these lines are undertaken by a number of teams [4, 5].

A second, more puzzling, failure mode exists. This is where the response message from the target object simply never arrives. What can the originating object make of this non-event? Did it receive the original message? Has the return message got lost? Should retransmission be the thing to do? In any event the application programmer should not be invited to get involved.

13.4.2.4 Concurrency transparency

As previously stated, the object model of the enterprise will be shared between many concurrent users. Application users will invoke operations on the objects that are public to them. During the execution of that operation, the object may invoke an operation on another object, which in turn will invoke an operation on a third object. Therefore, from a simple action by an application user, a stack of invoked objects will be built up. If a second application user were to perform an action that results in the middle object of this stack to be accessed and the second user were allowed access prior to the first user completing their transaction, then an inconsistent situation can arise.

Locking can be performed on objects to prevent the chance of the inconsistency being introduced. When the system uses object locking, the well-known deadlock situation can arise. In the example the second user will have to wait until the middle object in the first user's stack becomes free. Deadlock will occur if the first user's stack is blocked waiting for access to an object in the second user's stack.

These issues are only described here as indications of the concurrency problems with which the application programmer does not wish to become involved. The application code should remain free of exception cases to cope with concurrency conflicts.

13.4.2.5   Code-evolution transparency

When a system is installed and working, it is guaranteed to be subject to bug-fixing and enhancement requests. It will be essential, if the model is to continue to reflect the enterprise, that it can change with the enterprise or lapse into irrelevance.

This is the most difficult and least addressed transparency issue of all. If code evolution is not handled transparently, then the burden is put on programmers and administrators to provide evolution mechanisms. Experience shows that code evolution mechanisms have components that are both mechanistic, leading to boredom and therefore prone to errors, and complex, leading to confusion and therefore equally prone to errors. The result being that systems gain high inertia and act as a brake on the ability of an enterprise to evolve.

On the assumption that an enterprise model has been implemented and is doing good service on a day-by-day basis, a new requirement arises whereby additional data has to be stored with each customer object. New code is written and tested in isolation. Then the new code has to be introduced into the live system and the existing live customer objects have to be migrated to the new format. While easy to express it is an area of computer science that receives little attention and yet consumes huge amounts of *ad hoc* software engineering effort.

13.4.3   Requirements from the administrator's point of view

Administrators form the group within each enterprise who perform the tasks which keep the hardware and software infrastructure going throughout their life cycles. For hardware, this includes purchasing equipment, configuring it, connecting it to existing equipment, monitoring its performance, arranging repairs and eventually withdrawing it from service. For software, this includes purchasing, loading, configuring and arranging access for users. Also any software failures are reported back to support groups and corrective action is taken. It is the desire of the administrators, or at least their employers, that the effort expended on maintenance is minimal. Real world experience is often the reverse, with this group being frantically overworked. Some facets of self-managing systems that would assist with effort minimization are now described.

13.4.3.1   Self-healing systems

What often happens when something goes wrong with software is that a stream of cryptic messages is dumped in a log file from an embedded library and service is ceased. The log file is then conveyed to remote but knowledgeable individuals who tell the administrators how to recover the system. After this process has been repeated a number of times the administrators build up their own knowledge of diagnostic procedures. These procedures work until the next release of software changes the side effects of the embedded libraries.

The previously described failure-transparency mechanisms allow systems to carry on if a single node fails. This does not avoid the need for diagnostic procedures, but reduces the urgency of recovering a particular system. Service can be continued from the back-up systems without administrator intervention.

13.4.3.2   Self-tuning systems

Given that the system provides object-location transparency, it can decide where to put the objects. Presumably it will do this in such a way as to achieve optimum performance with a knowledge of the resources available. If objects have the ability to migrate at run time, they can move closer to the point where they are most often used. In an extreme case, the object could move on to the same network, into the same computer and even into the same process.

13.4.3.3   Self-installing systems

Application software usually needs to be installed on the targets on which it is to run. However, a system that holds a repository of executable code can be imagined. Whenever a new computer is added to the network, the system discovers its capabilities in terms of CPU, memory and disk space and downloads whichever executables it considers appropriate, perhaps using something like the UNIX facility 'ftp'. From that point onwards the computer can be a host to objects for which it has appropriate code.

13.4.3.4   Requests help when required

An object system that can perform all of the above tasks is inevitably undertaking a lot of resource monitoring. If the resources it has available are not sufficient to meet optimum performance, then an automated report can be made to the administrators. As a result, the administrators may enhance the configuration of the computing fabric.

13.5   THREE-LAYER STACK APPROACH TO SOLUTIONS

This section looks at possible solutions and considers how existing and emerging technologies may help. One of the most productive habits that developers can adopt is to think in terms of layers. If the approach is too rigid and many layers are artifically introduced, the resultant system becomes laden with inter-layer mapping libraries. Accordingly this chapter identifies only three layers, which can be named application layer, middleware layer and network layer, and are illustrated in Fig. 13.3.

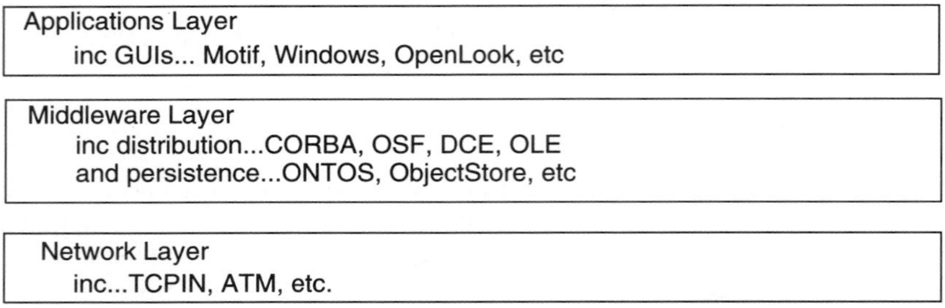

```
┌─────────────────────────────────────────────────────────┐
│ Applications Layer                                       │
│     inc GUIs... Motif, Windows, OpenLook, etc            │
└─────────────────────────────────────────────────────────┘

┌─────────────────────────────────────────────────────────┐
│ Middleware Layer                                         │
│     inc distribution...CORBA, OSF, DCE, OLE              │
│     and persistence...ONTOS, ObjectStore, etc            │
└─────────────────────────────────────────────────────────┘

┌─────────────────────────────────────────────────────────┐
│   Network Layer                                          │
│     inc...TCPIN, ATM, etc.                               │
└─────────────────────────────────────────────────────────┘
```

Fig. 13.3   The three-layer stack.

13.5.1   Existing network layer components

It is not the intention of this chapter to review in detail the network technologies and products available to implementors of enterprise models, but rather to underline the shift in design criteria for higher layers of the software. Availability of high-bandwidth, low-delay networks such as ATM-based systems means that programmers need not worry as much about making savings in these areas. A failure to recognize these shifts will lead to false design decisions being made in the name of performance, which will ease problems that are not bottlenecks.

13.5.2   Existing middleware layer components

Many products that address the requirements expressed in section 13.4 currently exist. The number of combinations in which these products can be connected to produce the required functionality is truly bewildering. Some of these products are considered below.

13.5.2.1   Operating systems

The operating system (OS) connects the computer software to the computer hardware and thus forms a foundation layer upon which other software is built. Previously, the relationship between hardware and operating systems was easily defined as both were supplied by the same vendor. Increasingly, the hardware and the operating system are becoming separately supplied and are becoming standard products.

UNIX™ is the most popular OS for workstations and is available on a wide range of machines from supercomputers to laptops. It is also widely used in distributed applications and many significant network innovations were made in UNIX and later ported to other operating systems. There are unfortunately competing variations of UNIX but the differences are reducing.

The latest OS offering from Microsoft™ [6] is called Windows NT and is a true multiprocessing micro-kernel based architecture. Importantly it exports a number of operating system applications programming interfaces (APIs) which map on to a set of base facilities. Among these APIs are Windows and POSIX, meaning it is an OS that can run both traditional PC and UNIX applications.

Although these two are likely to be dominant in the future, other options cannot be ignored. For instance, the trailing OS/2 from IBM now seems to be making a comeback as a contender.

13.5.2.2 Object databases

Object persistence, meaning the ability to store objects on disk and bring them into memory on an as-needed basis, is not a requirement in itself. The requirement is to support large shared-object models and object persistence is usually the mechanism by which the largeness criteria is satisfied. If another method arises that uses distributed shared memory in support of large object models, then disk-based systems may become unnecessary.

Object databases mostly operate by keeping a formatted copy of the in-memory object on disk. When an application needs to access the object, the object database system will copy it into the applications address space. Each object in an object database needs its own unique name that is valid whether the object is on disk or in memory. In order to create truly large object systems a number of object databases may have to be connected together. They then have to agree on consistent object names in the wider scope. This might cause problems as object databases may be used in conjunction with object request brokers (ORBs) which have their own object-naming requirements.

Object databases are not generally considered to be sufficiently mature as yet to support mission-critical applications. Also they do not yet have the built-in querying mechanisms that have made relational systems widely acceptable.

13.5.2.3 On-line transaction processing systems

On-line transaction processing (OLTP) systems have become part of the fabric of commercial programming although their need is not always immediately recognized. A system could be written and working for weeks before a crash occurs and the data is found to be tangled. This can occur when two applications both manage to edit parts of a data structure concurrently and leave the model in an inconsistent state.

The term 'transaction' has no direct translation into object model terminology. It is traditionally an application-level decision as to what collection of actions represents a transaction. Importantly the transaction is a collection of actions that may either be accepted as a whole or denied as a whole, leaving, in either state, the data in a consistent manner. This is done by locking out other applications from the data until the transaction is complete.

If the applications of the future are customizations of generalized object browsers then it would be useful if some automatic method could be found to define application transactions. All activity in the object model is initiated by the application user invoking top-level operations on public objects. Perhaps each such top-level operation is a transaction which locks out other applications until the result is delivered back to the application user. Transaction monitor code could be built into other middleware components such as object databases but currently OLTP must exist as a separate system. Use of existing OLTP systems involves extra work for programmers at development time and a significant number of extra processes at run time.

13.5.2.4   Distribution mechanisms

Mechanisms to provide client/server programming have been receiving a lot of attention as the benefits of distributed computing have become more obvious. The term client/server refers to an architecture where the application user interacts with a computer which does some local processing but then submits a request to another computer that performs more processing with access to more centralized resources and returns the result to the requester. The application user end is called the client and the other end is the server. The client/server relationship is therefore defined by the direction of the question/answer flow.

In a distributed object model the application user will invoke an operation on an object and if that result is returned immediately then the client/server relationship can be applied. However, most often that first object will want to invoke another operation on a second object in which case the original object becomes the client and the second object the server. To add further confusion the second object could, quite legitimately, call back to the original object to obtain further information, in which case the original object becomes both server and client at once. A better approach would be to say that each object was the equal or peer of the other and so no client/server relationship exists between them. Peer-to-peer architectures generally refer to those where any computer process can communicate directly with any other in order to initiate dialogues, request data and so are better suited to distributed object systems [7]. Accordingly, systems that have built-in client/server assumptions will not lend themselves so readily to object modelling. Some leading distribution mechanisms are considered below.

- OSF DCE and OODCE

  The Open Software Foundation (OSF) distributed computing environment (DCE) [8] is an integrated set of services that has been adopted by most UNIX suppliers and is primarily suitable for running in TCP/IP environments [7]. While its increasingly wide, and in some cases free, availability and standardization make it attractive to developers the functionality it provides is no more than has been available on bundled proprietary systems. However, it is another layer on top of the basic operating systems and so potentially serves to fill part of the middleware layer. The services that OSF DCE currently provide include a standardized remote procedure call (RPC) mechanism, security services, a useful directory service which is mostly used to locate servers, a 'threads' library for use in servers, and a distributed time service. OSF DCE can be considered a toolbox to help provide support for distributed object models. The toolkit is of too low a level to perform that task by itself. A Hewlett Packard (HP) product called OODCE extends the use of OSF DCE to support C++ objects and its definition is being submitted to the Object Management Group (OMG) [9] as a candidate for standardized adoption.

- CORBA

  Common Object Request Broker Architecture (CORBA) [9] is an architecture defined by the OMG which is a cross-industry consortia with some 400 members. CORBA contains a definition by which distributed object systems can be built. Current CORBA implementations provide host independence and some location transparency. The major co-developers of this standard are HP, DEC and Sun, all of which have their own offerings. The early lack of definition in the CORBA standard means that offerings from different suppliers will not interoperate, but this is expected to be fixed for future product releases. Other transparencies such as failure transparency are also expected to be introduced, but not in the near term. This activity has weighty industry support and holds genuine potential as a tool to support distributed objects.

  ORBs have a better approach than object databases with respect to the instantiation of objects. With an object database each application that wants to execute an operation on an object will get a copy of that object in its own address space. This can lead to one object copy per application invocation. The concurrency problems and duplicated resource usage to which this leads must be borne by the object database management system. The ORB takes a wiser approach for large object systems. The operation invocation is sent to the object wherever it currently resides. A combination of ORB and object database systems may be an approach for widely dispersed object models.

* OLE

  Microsoft [6] support their own distributed object model called Object Linking and Embedding (OLE). Their strength in the desktop market-place has led to attempts to allow the interworking of OLE and CORBA-based applications. Microsoft and DEC are collaborating on a product called the Common Object Model (COM) that allows CORBA objects and OLE objects to co-operate.

13.5.3   Existing applications layer components

As previously argued, the role of applications in a shared object environment can be reduced to a generalized object editor or browser. Each application would merely be a different restriction of the view of the object model based on who was using the browser.

Graphical user interface toolkits are often tied to their supplied operating system. However, OSF Motif [8] seems to be becoming the standard GUI provided with UNIX workstations and is available on virtually every client platform. This even includes the platform from Sun which has developed a serious competitor to Motif called OpenLook [10]. Apple, Microsoft and IBM (Presentation Manager on OS/2) all provide their own GUI programmer environments.

Browsers are graphical front ends to the object model and can replace the need for writing specific applications if they are powerful enough. Object database products such as ONTOS [11] and ObjectStore [12] are supplied with built-in browsers as are many relational systems.

13.6   A UNIFIED APPROACH TO MIDDLEWARE

The discussion so far has established the desirability and requirements of large shared-object models. This was followed by a description of the current and envisaged products that could form parts of the jigsaw. The problem is that in most cases they are parts that were originally designed to fit into different jigsaws. Fitting a selection of these components together from the existing choice will undoubtedly fill the middleware gap, but at what cost?

In order to produce a complete system an architect would have to pick an interworking collection of products and then build a team with the appropriate skills. Not only would these products need to be understood in isolation but the interactions and interfaces between them would need to be understood.

This would divert effort from the real issues of defining the enterprise model into the details and constraints of various packages.

The real problem is that the packages are mostly not specifically written to support large shared-object models but are enhanced or evolved to perform this task, which means they bring excess software with them. A fresh look at the problem would bring different and probably more elegant solutions. If the system is purely to support objects, then some of the necessary facilities can be built into those objects. The following paragraphs describe some ways in which the objects themselves could be empowered so they can contribute to large shared systems.

13.6.1   Empowerment of the objects with embedded performance tuning

Objects could for themselves keep track of which other objects are invoking their operations. In the case of a customer object which is resident on a computer in Hull and is constantly being accessed from computers in Devon each object could keep a table of its last 20 operations and decide whether to migrate to a computer closer to the invoking computer. In the extreme it could migrate on to the same computer and into the same process as the invoking object. The performance could then approach local performance. This is an example of how automatic performance tuning could be built into the middleware and hence avoid the need for human-driven tuning applications.

If the middleware supports cloned objects at a physical level, the tuning options become greater and more complex. The cloning of objects is primarily for resilience and requires that clones reside on different hosts. This acts against the interests of performance tuning and may lead to some interesting solution optimizations.

13.6.2   Empowerment of the objects with embedded transaction locking

Currently transaction locking is performed by external processes that monitor the progress of applications and enable 'roll back' or 'commit' as necessary. A more elegant approach would be to build an understanding of the need for transaction locking into each object. The object would 'know' not to commit the changes that an operation invocation may have produced until it received a message from the top-level object that the whole operation had been successfully performed. The object would also know how to roll back to the pre-invocation state if the transaction as a whole had failed.

13.6.3   Empowerment of the objects with the ability to display

Each object could collaborate with the browsers that are trying to display them by supporting an operation such as 'display' which when invoked would cause the object to make an image of itself appear on an appropriate screen. The object would need to have sufficient knowledge of its type and the security access that the invoker is allowed. Given this knowledge, general-purpose code could be written to present the object using the local GUI package.

13.6.4   Empowerment of the objects with failure recovery

It is difficult to see how the object that fails can arrange to recover without the application programmer, or even end user, having to intervene. If the object system employs some object replication, also known as 'groups', then a scheme becomes possible. With groups only one object exists from the programmer's and end-user's points of view; however, the implementor of the middleware will map this appearance to a number of physical objects. Each object should be an exact copy of the other in which case care must be taken that any edits to one of the groups are reflected to the others. In the event of a failure one member of the group becomes unavailable. The survivors of the group, who will be physically remote from the failure, could detect the absence of their erstwhile member. Recovery would be undertaken which would include generating another replica object on another trusted computer.

13.7   CONCLUSIONS

This chapter has indicated that large shared-object systems are likely to be the dominant software technology for IN applications and for enterprises in general. For use in an IN arena the object system needs to be highly available, widely distributed and have near real-time performance.

As computing and network infrastructure increases in performance while dropping in price the software design criteria emphasis must shift away from speed to consistent modelling and evolvability. What will hold back enterprises of the future is their inability to evolve their software safely and quickly. A successful large shared-object system will also support some form of code evolution.

Building the power to achieve the required functionality into the objects themselves appears to be the most promising approach but may lead to some conflicts. For instance, the methods proposed to achieve concurrency transparency may conflict with the mechanism for location transparency. For this reason a holistic approach needs to be adopted which hopefully will result in a unified elegant solution.

REFERENCES

1.   Meyer B: 'Object oriented software construction', Prentice Hall (1988).

2.   Cusack E L and Cordingley E S (Eds): 'Object oriented techniques in telecommunications', Chapman & Hall (1995).

3.   Lewis T J: 'Where is computing headed?' IEEE Computing Journal (August 1994).

4.   Rees R T O: 'The ANSA computational model', AR.001, APM, Poseidon House, Castle Park, Cambridge CB3 0RD, England (1993).

5.   Birman K P: 'The process group approach to reliable distributed computing', Communications of the ACM (December 1993).

6.   Microsoft Corporation, One Microsoft Way, Redmond, WA, USA 98052-6399.

7.   Elbert B and Martyna B: 'Client server computing', Artech House (1994).

8.   OSF: Open Software Foundation, Cambridge Centre, Cambridge, MA 02142, USA.

9.   OMG: Object Management Group, Inc., Framlingham Corp Centre, 492 Old Connecticut Path, Framlingham, MA 01701, USA.

10.   Open Look, XView Reference Manual, O'Reilly and Associates, Inc.

11.   ONTOS DBMS, Ontologic (1992).

12.   ObjectStore, Object Design Inc., 25 Mall Road, Burlington, MA 01803-4194, USA.

14

THE INFORMATION SERVICES SUPERMARKET — AN INFORMATION NETWORK PROTOTYPE

I W Marshall and M Bagley

14.1 INTRODUCTION

The intelligent network (IN) described in earlier chapters of this book has a vision outlined in the International Telecommunications Union (ITU) standards [1] as follows:

> 'The objective of IN is to allow the inclusion of additional capabilities to facilitate provisioning of service, independent of the service/network implementation in a multivendor environment. Service implementation independence allows service providers to define their own services independent of service-specific developments by equipment vendors.
>
> Network implementation independence allows network operators to allocate functionality and resources within their networks and to efficiently manage their networks independent of network implementation-specific developments by equipment vendors.'

A standard which fully supports this broad vision would certainly satisfy many of the requirements of telecommunications service providers, well into the 21st century. However, by the time the Q.1200 series of IN standards was published in 1992 it was clear that they only supported a small part of the overall vision. It was also clear that further progress within the normal standards process would be very slow.

At the same time it became apparent that there was considerable convergence between the computing industry and the telecommunications industry. The common vision of both the telecommunications operators and the computer industry was of a future based on information networking. An information network is an infrastructure which:

- facilitates business processes and the efficient use and reuse of information;

- allows diverse businesses to co-operate in whatever way they choose;

- supports user-customizable interactive multimedia applications;

- allows seamless interworking between distributed information sources.

Clearly, this infrastructure will need service implementation independence to be able to deal with the enormous diversity of sources, applications and hosts. It will also need strong standards for interworking — in other words it needs the IN vision to be fully realized.

As part of the long-term IN development process many telecommunications operators had been informally collaborating on this information networking vision through a series of workshops, known as telecommunications information network architecture (TINA) workshops since 1990. At these workshops it was identified that the convergence was both a threat and an opportunity — a threat because it increased the competition and made the need for realization of the IN vision more urgent; an opportunity because it seemed that the computer industry had some of the answers and was willing to share its solutions in order to ensure a wide market base.

Managers are thus presented with a triptych of problem, threat and opportunity. One proposal to facilitate the solution of the problems, and thus minimize the threat and make use of the opportunity, was to establish a formal collaboration between the telecommunications operators, the computer vendors and the network equipment vendors which would attempt to bypass the policital wrangling associated with formal standardisation and produce a quick *de facto* standard. This collaboration goes by the name of the TINA Consortium (TINA-C) [2].

This chapter outlines the problems with the IN standards, with respect to the information networking vision, and then introduces TINA-C and its approach to resolving the problems. It then illustrates the use of TINA-C by discussing the implementation of a possible future telecommunications service retailer, and the benefits to the retailer of using the TINA-C approach; this work is being conducted in a project being undertaken at BT Laboratories called ComBAT (common broadband applications testbed).

14.2   THE INFORMATION NETWORKING VISION

Traditional telecommunications operators' markets are now highly competitive, and BT, along with many other telecommunications operators, is responding by seeking new markets and engaging in new activities to serve them. One such market-place is information networking. A selection of information networking services is listed in Table 14.1.

Table 14.1   Examples of multimedia information services.

Interactive services (symmetrical duplex communication)	Distribution services (simplex communication)
Multimedia Tele-education Tele-shopping Building security Traffic monitoring LAN-to-LAN interconnect Medical imaging Remote control Computer-supported co-operative working Electronic mail Document transfer Remote control Video on demand Electronic library	Share prices Weather Electronic newspaper Electronic publishing Tele-advertising Tele-education

Until recently, the principal barriers to mass-market introduction of commercial information services of this kind have been technological and financial. The Internet, an interconnected user group bound by their interest in exploring, sharing and using multimedia applications, is experiencing explosive growth fuelled by the ready availability of easy-to-use cheap terminals. The range and sophistication of the services available on the Internet is impressive and most of them are free, and many are now available via user-friendly hypertext interfaces. It might appear that the principal barriers have been overcome.

However, the services are currently not well commercialized, the user is not sure what to expect from a service, it is not guaranteed to work, a user cannot request exclusive or limited use of the service to improve quality and finding the location of a particular service is by word of mouth. The Internet, as the best known example of a primitive information network, could be likened to an information bazaar. The buyer has to beware of false claims, overcharging, fly-by-night suppliers, etc, and where successful linking of the user and the desired information can be extremely time consuming.

At present commercial information services, with more guarantees to the user, are starting to appear. Examples include Commerce-net (via World Wide Web), Prodigy (supplied by IBM), specialist data providers (e.g. Mead Data) and video-on-demand services (e.g. BT). These services could be compared to an information high street — good services are available but you have to walk all over town to find them and write dozens of cheques.

The ultimate aim is to provide an information hypermarket (or supermarket) which stocks the users' needs on its shelves (whether they are ready-made applications or pick-and-mix features, own label or third party), provides easy access, has a single billing point and can be seen as a trusted supplier. At current rates of progress this vision will be realized well before the end of the century.

Such an information network will be required to be:

- ubiquitous (to ensure easy access);

- dependable;

- capable of providing user-defined quality of service;

- secure;

- interoperable;

- easy to use;

- responsive;

- anything else a user requires/demands.

To achieve all the above for the complex sophisticated services envisaged it is not enough to provide a broadband network and some simple services that run over it (such as Freefone and video links). This would be analogous to providing the road network and the fleet of lorries (with cab radios) which are used to ensure the hypermarket is fully stocked. A wide range of management tools and advanced processing techniques are also required. Many of these can be borrowed from the computer industry and built on to the IN infrastructures currently being deployed, but firstly the problems with existing standards must be resolved and interoperability fully understood.

14.3   PROBLEMS WITH INTELLIGENT NETWORK STANDARDS FOR INFORMATION NETWORKS

14.3.1   The problem of scope

IN is claimed to be applicable to public switched telephone networks (PSTN), public data networks (PDN), integrated services digital networks (ISDN), and all the services defined for these networks. There are two major difficulties:

- the standards only contain protocols for usage of the PSTN and ISDN and do not address the integration of local access network (LAN) technology or PDN technology;

- the services defined in the standards cover only traditional telecommunications company services such as virtual private networks (VPN), universal personal telecommunications (UPT) and Freefone — they do not currently address the needs of information access services (such as inter-process negotiation on requirements and presentation) at all.

14.3.2   The problem of interworking

The IN standards attempt to define 'standardized communication between network functions via service-independent interfaces'. However, the standards were not originally developed with the intention of supporting interworking between telcos, but only between the suppliers to an individual telco. As a result, the standards work well as a procurement template, but do not contain enough flexibility to allow inconsistent implementations which do not interwork (e.g. no global Freefone). This is partly due to a lack of detail in the prescriptions, and partly due to the scope of the definitions, which only attempt to standardize communication between network functions, and do not address communication between network applications. The first problem is being addressed (UPT and global system for mobility (GSM), see Chapter 2), but due to associated political sensitivities the solution will consume considerable effort. This means that very little progress is being made on solving the second problem.

14.3.3   The divergent standards problem

The IN standards cover the domains of network and service management; however the Telecommunications Management Network (TMN) standards [3] (which are derived from the open systems interconnect (OSI) model) deal with management issues in a different way (which is used in ISDN and synchronous digital hierarchy (SDH)). It is not at all clear how these standards are intended to be used together. In addition, new standards such as open distributed processing (ODP), and its derivatives, appear to be likely to increase the confusion further, since they assume a radically different origination of software, not used in any existing standards.

14.3.4   The problem of time

Typically a new standard takes four years to become a stable committee draft and gain approval. The improvements to the IN standards currently under way will be no exception, and work on some of the problems has not even started.

14.4   TINA-C

TINA-C is a consortium of approximately 40 telecommunications companies, computer vendors and network equipment vendors, which commenced work in 1993. The telecommunications companies' representatives include such major players as BT, AT&T, Bellcore, NTT, DBP and Cable and Wireless; the network equipment vendors include such major players as Ericsson, GPT and Northern Telecom; the computer vendors include such major players as IBM, HP and DEC. The organization consists of a core team of 40 researchers, based at Bellcore Labs in New Jersey, USA, and a set of auxiliary projects run independently by the partner companies. TINA-C is defining a *de facto* standard information networking architecture, which will guarantee interoperability between information networks designed using the architecture. The complete architecture will be available some time in 1997. It will be based as much as possible on existing standards and public domain results [2, 4, 5] and will promote:

- universal information access and transport;
- rapid service provisioning;
- tailored services;
- third-party development of services and applications;
- service security, reliability and performance.

The process involves the core team describing, the auxiliary projects implementing and criticizing, and the core team prescribing on the basis of the results. To date the published results are descriptive, but substantial input is being provided by the auxiliary projects which should lead to preliminary prescriptive results early next year (i.e. after two years rather than the usual four for a formal standard). The principal agreements and descriptions are summarized below.

The TINA architecture is based on applying ODP principles of object orientation and distribution to telecommunications system design using TMN-managed objects and IN concepts for service management and control.

The TINA-C architecture is based around a core of an object management group (OMG) [6] distributed processing environment (DPE). This provides for communications between objects, dynamic bindings via a trader function, and notification servers to provide management information (e.g. faults, performance). Extensive use is made of autonomous collections of distributed objects referred to as agents [7-9]. The architecture is presented in a set of deliverables which are written in a mixture of English and formal object-oriented representations such as Rumbaugh [10] and Guidelines for the Definition of Managed Objects [3].

At the highest level, the TINA-C entities resident on the DPE can be summarized by Fig. 14.1, which describes the various components TINA-C has

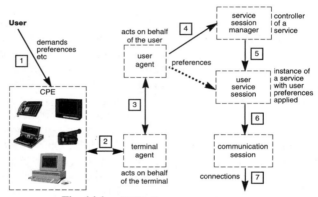

Fig. 14.1   TINA-C access components.

defined to be necessary to set up a service connection. As the components described are recursive, if a service was made up of several other (more basic) services, then each service component could potentially have an associated service session for management of that component.

These TINA components and the interactions between them are now described.

- A user makes some demands (Fig. 14.1, interaction 1) via the set-top box, PC or telephone, i.e. the customer premises equipment. The user can make inputs to impose some personal (or global) preferences either directly (via a keyboard, keypad, swipe card, etc) or indirectly by pointing at preferences for other services.

- Any changes and demands are passed on (Fig. 14.1, interaction 2) to the user's terminal agent (situated locally, but could be held remotely), which can then act on some of the demands (or preferences) and add information to the demands (for instance CPE capabilities).

- The terminal agent then passes on (Fig. 14.1, interaction 3) any information directed to the user agent, along with its additions. The link from the user's CPE to the user agent via the terminal agent is a logical one, the terminal agent merely acting as a filter, adding or removing information as required.

- The agent acting on behalf of the user (the user agent) then makes requests (Fig. 14.1, interaction 4) to the relevant service session manager (say for a video conferencing service) about the requirements that the user has directly or indirectly expressed (through preferences communicated via the terminal agent to the user agent).

- These requests are acted on by a factory (Fig. 14.1, interaction 5), which is a component of the service session manager. The factory produces a customized version of the service session, called a user service session. Depending on the constraints imposed by the service session the user may now have some level of control over the service session. For the example of a video conference, the service session class may be set up in such a way as to give chairman privileges to the first entrant into the video conference (or the first user of the video conference service session factory object). The user agent, or the user, now has to inform the other participants in the conference that they are in a conference in progress. In the case of other users, they too will have a measure of customizability and control over the conference session since they will have their own user service session interface. This interface allows them to define the bandwidth of video windows that are not in use (for instance) and define what video streams should go the set-top box, which video or audio streams to store via the PC, etc. They will, however, only have limited control over the general controls of the conference, how to gain the floor, etc.

- The user service session will inherit certain functionality associated with a conference, the need for video and audio streams and a data stream, all potentially duplex and separately manageable. This is achieved by creating (via a factory) objects called connection sessions that are managed by a communication session (Fig. 14.1, interaction 6). This delivers the potential to dynamically configure the streams in a conference, either directly (i.e. choose not to send video) or as part of the service (reduced bandwidth allocation to a channel that is not directly involved in the conference) or imposed by the user agent (take advantage of low tariff periods, etc) or the terminal agent (terminal capacity constraints).

- The communication session sets up connection sessions (Fig. 14.1, interaction 7) as required (including packet-switched virtual connections for file transfer, etc).

Figure 14.2 shows how the connections for a videoconference can be picked and mixed; in this case the conference is made up of separate objects controlling asynchronous transfer mode (ATM) and ISDN audio and video streams.

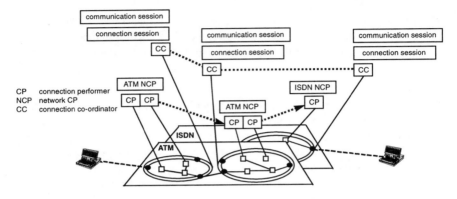

Fig. 14.2   Connection management.

14.5   INFORMATION SERVICES ENTERPRISE MODEL

The BT ComBAT project is attempting to realize an optimal implementation of an information networking environment. ComBAT is primarily focused on the service provision and management aspects not covered by network development or application development. To express the domain of interest more clearly an enterprise model (or operations scenario) has been developed. The enterprise model was derived by interpreting information networking as information retailing and attempting to represent the activities of a hardware retailer/ distributor [11, 12] in the model. This analogy is the source of the name information service supermarket (or hypermarket). It was felt that basing the model on an existing implementation in this way would enable productive reuse of existing technology and knowledge, and thus minimize the effort involved. Analysis of this enterprise model has led to the view that a distributed object-oriented implementation with well-defined open interfaces is needed. Many (but not all) of the concerns are addressed by TINA-C, so TINA-C has been used as the basis of the designs. The model is shown in Fig. 14.3.

There are four separate areas depicted in the model:

- the user domain;
- the network domain;

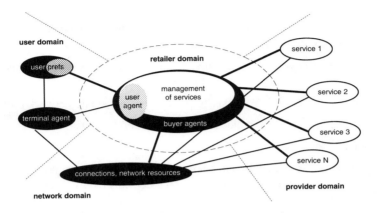

Fig. 14.3   Information services supermarket.

- the provider domain;

- the retailer domain.

The user domain represents the domain of interest of users of the services available in the retail domain. It contains the users and items which they own or control such as their personal profiles and applications, and their terminal(s). It also contains a terminal agent which can negotiate with the retailer domain when necessary. This agent is intended to include the terminal agent concepts from the TINA-C architecture. It is envisaged that the terminal agent may be supplied by the retailer but owned by the user (or their subscriber).

The network domain represents the domain of interest of the owners of the network used to convey messages from the user to the retailer or other users. It is envisaged that by analogy with 'normal' retailing the user will not have a direct contractual relationship with the network. Instead the network should be regarded as a subcontractor to the retailer. Initial access of the user to the retailer is regarded as being achieved by services such as Freefone which are paid for by the retailer. The network domain is intended to contain the network hardware and associated software, such as connection co-ordinators and managers from the TINA connection management architecture. It also needs to contain connection control features from IN capability set 1/2 (CS-1/2) to enable Freefone and mobility.

The provider domain represents the domain of interest of sources of the services which can be purchased in the retail domain. It contains a set of disparate service providers, each of which provides agents that can negotiate with buyers in the retail domain on supply and payment for services, and similar issues. It is probably a specialization of the retailer domain with a restricted set of customers and service offerings. This allows the purchase of composite services.

The retailer domain is perhaps the most interesting. It is intended to represent the domain of interest of a retailer who can:

- facilitate access to information and communications services and associated tools;

- act as a middle man or broker for service providers and network providers;

- offer customized, guaranteed services to individual customers;

- manage the services.

It contains a set of intelligent agents which can respond to user and manager demands on an individual basis. These include the TINA-C user agent, service session manager and communications session manager. It also contains a set of management processes which allow managers to:

- monitor activity;

- make configuration changes;

- specify and enforce policies;

- maintain quality, performance, access, etc, as demand and resources vary;

- enforce payment and security-related mechanisms;

- assess interactions and add links to third parties;

- advertise services.

A more complete list can be obtained by analysing the activities of existing retailers. It has been found that the above services can be supported by the existing TINA-C resource, fault and configuration concepts, and that the concepts can be reused for high-level designs of security, account and performance management processes. This retailer can be considered almost entirely analogous to the arrangement a supermarket (say a food retail one) has with suppliers of products and the services it offers to those suppliers and the buyers of the products (subscribers). Services (products) are offered by the supermarket on behalf of the service providers to users (end-users that are associated with a subscriber). The service offered can in theory be physically situated anywhere, so it could be resident in the provider's domain and accessed via the supermarket, or it could be resident in the supermarket. In practice, the latter is often the more convenient position for it, since the cost of large servers supporting many users can be shared between a range of services. On the other hand it may be that the provider wishes to retain the management of their service and merely devolve the access control (thus sharing security costs), or enable their service to be located by an intelligent navigation service offered by the retailer, in which case the server will be in the provider domain.

The model illustrates several important inter-domain interfaces. These include:

- the user domain/retailer domain interface;
- the retailer domain/provider domain interface;
- the retailer domain/network domain interface;
- the user domain/provider domain interface (mediated by the retailer).

In order to support flexible use of services from different providers and connections across different networks, making use of user-domain terminals and applications of diverse types, these interfaces need to be open, i.e. the capabilities required at the interface can be described using a set of defined terms drawn from published standards and agreements. A key concern is to establish the degree of openness required at the inter-domain interfaces and agree how it can be achieved. Refining the descriptive specifications currently provided by TINA-C into a prescriptive standard relies on input from partners focused on the roles they might play in the enterprise model presented earlier. The understanding of the interfaces and their properties gained from the design work, and the degree to which TINA-C enables them to be open, is discussed as an illustrative example in the final part of this chapter.

14.6   SUPERMARKET DESIGN

As an example, an information services supermarket service-access scenario will be used to illustrate a realization of a system within the model described above. Figure 14.4 shows a refinement of the enterprise model which illustrates the classes of components used in the design. The scenario contains a single service provider, a single network provider, a single retailer and a set of users.

Providers use their applications (authoring tools) to create or extend a service. The service is wrapped with some additional code in order to allow it to function in the supermarket (retailer). The wrapper is defined following a negotiation between the agents in the provider domain and the agents in the retailer domain, and its precise form will depend on the relationship between the provider and the retailer. Retailer applications are then used to manage the access of customers of the supermarket to the service. Information about the retailer's customers (subscribers) and users are held by the supermarket in support of, for example, user customization of services and global user preferences.

Users (no matter what role they are playing) access the system via a terminal of some kind (say a personal computer or a set-top box of comparable processing power, with some technology to access the system, DPE kernel, network card, code to access the services on the system, etc). The user initially calls up the supermarket via some kind of Freefone service, which the user has identified

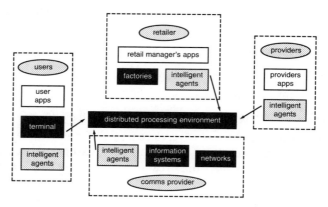

Fig. 14.4   Information network framework.

through some form of advertising (possibly via the network provider to which the user is connected). On login, authentication takes place; on successful authentication, user agents are provided to the user which give them certain capabilities depending on what a subscriber has provided for them.

The agent used will depend on the role that the user is playing. A number of agents and a mechanism for transforming from one role to another have been defined and designed. The agents and their relationships are depicted in Fig. 14.5.

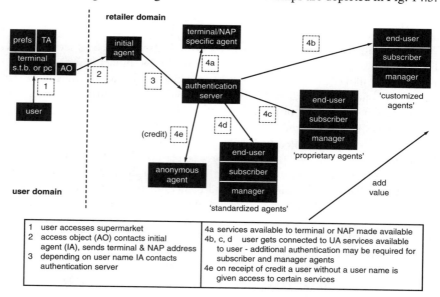

Fig. 14.5   User agents.

As can be seen from Fig. 14.5, a number of agents are envisaged, the detailed functions of which are described below, along with some other agents not depicted in Fig. 14.5:

- sales agents — these are the agents the users see when they have logged on to the supermarket and wish to browse and/or try out services; they need to supply subscriber authentication to buy (get access to) services;

- subscription agent — an agent for the purchase of services for use by end-users, acting for a number of end-users;

- management agent — an agent for the management of the service, i.e. service operator role;

- service provider agent — the access the service provider has to the service provider's service logic;

- buyer agent — an agent for the purchase of service provider services for resale/inclusion in another service, the actions being split between those by means of which the buyer can act upon the service provider and those that can act on the supermarket;

- initial agent — the initial agent that is available on start-up of a service for the first time; control is passed to any of the other agents — to which one depends on the actions of the user in identifying to the supermarket the role being played;

- terminal/NAP agent — services that have been made available to the terminal or network access point (NAP) are used via this agent;

- end-user agent — the end-user aspects; the end-user is not the provider (i.e. subscriber) of the service and therefore cannot browse.

Once the users have obtained their agents they can request any of the services the agents authorize them to access by passing their requests to a factory. The factory first creates a generic service session of the type requested by running the appropriate set of components from a library. The service session then adds any additional user-specific components which are required and becomes a user (specialized) service session. The specialized session then makes use of communications sessions, as in the TINA architecture. Figure 14.6 illustrates these interactions.

The design is a refinement of the TINA architecture in a number of important respects including:

- the access agents' explicit support for authorization and role authentication;

- the support for user roles;

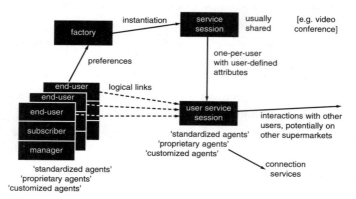

Fig. 14.6   Session interactions.

- the user service session not necessarily being a standardized component;

- the service session being inclusive of the run-time service session management.

Many of the refinements are now being adopted by the TINA-C core team for inclusion in future deliverables.

Concurrent implementation experiments of this design (in the true spirit of object-oriented design) are being performed at BT Laboratories. The experiments assumed an ATM network platform since this could provide all the network properties the scenario demanded, without involving substantial network integration tasks. The proprietary signalling scheme supplied by the switch manufacturer was used, since the open ATM standards are not yet mature. Off-line network hardware management (diagnosis, planning, etc) is performed with the assistance of HP Openview. The DPE was based on the OMG compliant Orbix platform supplied by Iona Technologies. A trader was added based on the implementation in Architecture Project Management's ANSAware. So far the implementations have not realized the flexibility inherent in the design since the interfaces are quite complex (see section 14.7) and there has not been time to implement all the features necessary. Details of the initial implementation experiments can be found elsewhere [13].

14.7   INTER-DOMAIN INTERFACES

In support of the operations of the supermarket there will be transfers of information across the inter-domain interfaces identified in the enterprise model. The interfaces must allow heterogeneity of applications, terminals, user interfaces, networks and servers. They must also support personalized user preferences, personal and terminal mobility, secure authentication, authorization and billing, etc. A full definition is clearly going to be complex; for example Table 14.2 lists many of the capabilities and concepts needed at the interface between the user domain and the retailer domain (i.e. between the terminal agent and the access agent), together with an indication of the wide range of standards which could be used to make the interface open.

It is clear from the table that TINA-C could be the source of many of the open interface concept definitions required. These experiments will help to steer the TINA-C team towards prescriptive definitions and accelerate progress to a strong standard. The experiments are also providing informed input to a wide range of development programmes.

Table 14.2   A listing of capabilities at the terminal agent/access agent interface, the concepts which must be defined to support them, and some appropriate standards from which the concepts can be drawn.

	Network and DPE		Terminals	Applications	User interfaces	Users
Required capabilities	i) Protocol translations	ii) Communications between heterogeneous DPEs iii) Definition of end points, etc (what is a network?)	Owner defined capabilities Resolve conflicts with user preferences Hardware-independent O.S.-independent	Version and manufacturer compatibility Portable Flexible configuration and deployment	Reproduce user defined look and feel on any device Choose my menus, VR, text styles, colours, speed, and still communicate with capabilities	Set filters Deal with incompatible filters Pick-and-mix features
Defined concepts needed to support requirements	i) Invocation Reply Notification Argument Result Object Interface Stream Source Sink Protocol	ii) Bindings Channel Definition Ordering Synchronization Transactions iii) G.803	Nucleus Wrappers Capabilities Properties of hardware (machine profiles)	Behaviour Dependencies Relationships Entity Format	Express underlying needs, and shared attributes Agree capabilities	Constraints: time location recipient price interaction QoS permissions details
Relevant standards	OMG TINA-C?	OMG IETF TINA-C? OSI	DAVIC SPIRIT TINA-C OMG	TINA-C Multimedia forum X-Open, OLE	Nextstep X-11, Microsoft Windows	TINA-C + auxiliary projects, selected from ITU

DAVIC   The Digital Audio-Visual Council
OSI   Open Systems Interconnect
OLE   Object Linking and Embedding (Microsoft)
SPIRIT   Service Provider Integrated Requirements for Information Technology

IETF   Internet Engineering Technical Forum
X,etc   Computer Video Display Standards
Nextstep OO Operating System

14.8   CONCLUSIONS AND RECOMMENDATIONS

Many globally aspiring telcos have a strategic business aim of being able to procure interactive multimedia services world-wide in a plug-and-play environment and so give the customer the ultimate in choice. Through an exemplary design, the understanding of the delivery of the advanced features and services currently being discussed has moved on to a stage where the problems appear solvable. A good architecture on which to build solutions and the required open interface standards is TINA-C.

TINA-C was set up in 1993 in an attempt to short cut the standardization process and build a standard for information networking based around the IN, TMN and ODP standards. To date, the consortium has delivered a set of concepts on which the standard will be based. It has proved possible to create a high-level design of an advanced information service and its environment using the TINA concepts, and then successfully implement large parts of that design using commercially available tools. However, major gaps in TINA have been identified, especially for user interface concerns, and much progress needs to be made in the standardization of key interfaces. In conclusion, while the set of concepts offered by TINA is useful, it needs to be extended to cover these gaps. Further developments should concentrate on the provision of an agreed high-level interface definition language which should be thought of as a client of the OMG interface definition language.

REFERENCES

1.   ITU-T recommendation Q.1201: 'Principles of intelligent network architecture', (1992).

2.   Chapman M: 'Overall architecture', TINA-C Deliverable (1994) [http://tinac.com:4070/0/94p2/viewable/overall.ps]

3.   Smith R et al (Ed): 'The management of telecommunications networks', Ellis-Horwood (1992).

4.   Berndt H and Minerva R (Eds): 'Definition of service architecture', TINA-C Deliverable (1994) [http://tinac.com:4070/0/94p2/viewable/servarch.ps]

5.   Kobayashi H (Ed): 'Service component specifications', TINA-C Deliverable (1994) [http://tinac.com:4070/0/94p2/viewable/servcomp.ps]

6. OMG: 'The common object request broker architecture and specification', Doc No 91.12.1, Revision 1.1, Draft 10 (December 1991).

7. Arango M et al: 'The touring machine system', Comm ACM, 36, No 1 (1994).

8. Kautz H, Selman B and Coen M: 'Bottom up design of software agents', Comm ACM, 37, No 7 (1994).

9. White J: 'Telescript technology: the foundation for the electronic market-place', General Magic White Paper (1994).

10. Rumbaugh J et al: 'Object oriented modelling and design', Prentice Hall (1991).

11. Darabi F and Howard-Healy M: 'Virtual private networks: market strategies', Ovum (1992).

12. Jeffcoate J and Templeton A: 'Multimedia strategies for the business market', Ovum (1992).

13. Bagley M, Marshall I W et al: 'The information services supermarket — a trial TINA-C design', TINA Conference, Melbourne Australia (February 1995).

Appendix —
List of acronyms

ACD	automatic call distribution
ACSE	association control service element
ACTS	Advanced Communications Technology and Services (a European initiative)
ADPN	administration data packet network
ADSI	analogue display services interface
AE	application entity
AEI	application entity invocation
AIN	advanced intelligent network
AIP	advanced information processing
AM	applications manager
AMPS	Advanced Mobile Phone Service
ANSI	American National Standards Institute
API	application programmers interface/application process invocation
ASE	application service element
ASN.1	abstract syntax notation 1
ATM	asynchronous transfer mode
AuC	authentication centre
BCP	basic call process
BCSM	basic call state machine (model)
B-ISDN	broadband ISDN
BSC	base station controller
BSS	base station system
BSSAP	base station system application part
BSSMAP	base-station system management application part
BTS	base transceiver station
CASE	computer-aided software engineering
CCAF	call control agent function

CCF	call control function
CLASS	customer-calling local area signalling system
CLI	calling line identity
CNA	co-operative networking architecture
COM	Common Object Model
ComBAT	common broadband applications testbed
CORBA	Common Object Request Broker Architecture
CPE	customer premises equipment
CPU	central processing unit
CS-1	capability set 1
CSS	customer service system
CSTA	Computer Supported Telephony Applications — the ECMA standard for CTI
CT2	Cordless Telephone 2
CTI	computer/telephony integration — this is the currently accepted acronym, though the concept has been referred to by a number of other terms, e.g. computer supported telephony (CST), computer integrated telephony (CIT), host control integration (HCI)
DAVIC	Digital Audio-Visual Council
DCE	distributed computing environment
DDSN	digital derived services network
DECT	digital European cordless telecommunications
DLI	digital line interface
DLL	dynamic link library
DP	detection point
DPE	distributed processing environment
DPNSS	digital private network signalling system
DSS1	Digital Subscriber System No 1
DTAP	direct transfer application part
DTI	Department of Trade and Industry
DTMF	dual tone multifrequency
EC	European Commission
ECMA	European Computer Manufacturers Association
EF	elementary function
EIR	equipment identity register
ETSI	European Telecommunications Standards Institute
FBB	functional building block
FCC	Federal Communications Commission
FDDI	fibre distributed data interface
FEA	functional entity action
FPLMTS	future public land mobile telecommunications systems
GMSC	gateway MSC
GSL	global service logic

GSM	global system for mobile communications
GUI	graphical user interface
GVNC	global virtual network communications
HLR	home location register
IETF	Internet Engineering Task Force
IN	intelligent network
INAP	intelligent network application part/protocol
INCM	intelligent network conceptual model
IP	intelligent peripheral/Internet protocol
ISDN	integrated services digital network
ISO	International Standards Organization
ISP	intermediate services part
ISUP	ISDN user part
IT	information technology
ITU	International Telecommunications Union
IWF	interworking function
LAN	local area network
LCP	local control point
LE	local exchange
MACF	multiple association control function
MAP	mobile application part
MPRN	multi-protocol router network
MS	mobile station
MSC	mobile switching centre
MT	mobile terminal
MTP	message transfer part
MVIP	multivendor integration protocol — a standard for multi-64 kbit/s channel interconnection between PC expansion boards, independent of the PC bus
NACC	network administration computer centre
NAP	network access point
NCP	network control point
NFS	network file system
N-ISDN	narrowband ISDN
NMT	Nordic Mobile Telephone
NOMS	network operations management system
NOU	network operations unit
NSP	network services part
NUP	network user part
ODP	open distributed processing
OLE	Object Linking and Embedding
OLO	other licensed operator
OLTP	on-line transaction processing

OMC	operations and maintenance centre
OMG	Object Management Group
OO	object oriented/object orientation
ORB	object request broker
OS	operating system
OSF	Open Software Foundation
OSI	open systems interconnection
OSS	operational support system
P(A)BX	private (automatic) branch exchange
PC	personal computer
PCM	pulse code modulation
PCMCIA	Personal Computer Memory Card International Association
PCS	personal communications services
PDA	personal digital assistant
PIC	point in call/personal intelligent communicator
PMMB	provided, maintained, monitored and billed
PN	personal numbering
POTS	plain old telephony service
PSTN	public switched telephone network
RBOC	Regional Bell Operating Company
RIL	radio interface layer
ROSE	remote operations service element
SACF	single association control function
SAO	single association object
SAP	speech applications platform
SAPI	service access point identifier
SCAI	Switch-Computer Applications Interface — the ANSI standard for CTI
SCCP	signalling connection control part
SCE	service creation environment
SCEF	service creation environment function
SCF	service control function
SCP	service control point
SCSA	Signal Computing System Architecture — an architecture for the integration of voice peripherals in a PC, which includes an API and the specification of a multi-64 kbit/s interconnection between expansion boards (although specified by Dialogic, a number of other vendors are now following this 'standard')
SCSI	small computer systems interface
SDF	service data function
SDL	Specification and Description Language
SDP	service data point
SIB	service-independent building block

SIM	subscriber identity module
SUR	service-independent usage record
SLEE	service logic execution environment
SLI	service logic interpreter
SLP	service logic program
SMAF	service management agent function
SMF	service manacement system
SN	service node
SNAC	service node accounting and charging
SPC	stored program control
SPIRIT	Service Provider Integrated Requirements for Information Technology
SRF	specialized resource funtion
SS7	Signalling System No 7
SSF	service switching function
SSP	service switching point
TACS	Total Access Communications System
TAPI	Telephony Applications Programming Interface (Microsoft)
TASC	Telecommunications Applications for Switches and Computers — the ITU standard for CTI
TC	transaction capability
TCAP	transaction capabilities application part
TCP/IP	transmission control protocol/Internet protocol
telco	telecommunications company
TINA	Telecommunications Information Network Architecture
TINA-C	TINA Consortium
TMN	Telecommunications Management Network
TSAPI	Telephony Services Applications Programming Interface (Novell)
TSPI	telephony service provider interface
TUP	telephony user part
TX	transit exchange
Um	air interface
UMTS	universal mobile telecommunications system
UPT	universal personal telecommunications
VLR	visitor location register
VoD	video on demand
VPN	virtual private network

Index